THE CALCULUS GALLERY

WILLIAM DUNHAM

THE CALCULUS GALLERY

Masterpieces from Newton to Lebesgue

PRINCETON UNIVERSITY PRESS

PRINCETON AND OXFORD

Copyright © 2005 by Princeton University Press
Published by Princeton University Press, 41 William Street,
Princeton, New Jersey 08540
In the United Kingdom: Princeton University Press, 3 Market Place,
Woodstock, Oxfordshire OX20 1SY

Library of Congress Cataloging-in-Publication Data
Dunham, William, 1947–
The calculus gallery : masterpieces from Newton to
Lebesgue / William Dunham.
p.   cm.
Includes bibliographical references and index.
ISBN 0-691-09565-5 (acid-free paper)

1. Calculus—History.   I. Title.
QA303.2.D86 2005
515—dc22
2004040125

British Library Cataloging-in-Publication Data is available

This book has been composed in Berkeley Book

Printed on acid-free paper. ∞

pup.princeton.edu

Printed in the United States of America

1   3   5   7   9   10   8   6   4   2

In memory of Norman Levine

Contents

Illustrations

Acknowledgments

This book is the product of my year as the Class of 1932 Research Professor at Muhlenberg College. I am grateful to Muhlenberg for this opportunity, as I am to those who supported me in my application: Tom Banchoff of Brown University, Don Bonar of Denison, Aparna Higgins of the University of Dayton, and Fred Rickey of West Point.

Once underway, my efforts received valuable assistance from computer wizard Bill Stevenson and from friends and colleagues in Muhlenberg's Department of Mathematical Sciences: George Benjamin, Dave Nelson, Elyn Rykken, Linda McGuire, Greg Cicconetti, Margaret Dodson, Clif Kussmaul, Linda Luckenbill, and the recently retired John Meyer, who believed in this project from the beginning.

This work was completed using the resources of Muhlenberg's Trexler Library, where the efforts of Tom Gaughan, Martha Stevenson, and Karen Gruber were so very helpful. I should mention as well my use of the excellent collections of the Fairchild-Martindale Library at Lehigh University and of the Fine Hall Library at Princeton.

Family members are a source of special encouragement in a job of this magnitude, and I send love and thanks to Brendan and Shannon, to my mother, to Ruth and Bob Evans, and to Carol Dunham in this regard.

I would be remiss not to acknowledge George Poe, Professor of French at the University of the South, whose detective work in tracking down obscure pictures would make Auguste Dupin envious. I am likewise indebted to Russell Howell of Westmont College, who proved once again that he could have been a great mathematics editor had he not become a great mathematics professor.

A number of individuals deserve recognition for turning my manuscript into a book. Among these are Alison Kalett, Dimitri Karetnikov, Carmina Alvarez, Beth Gallagher, Gail Schmitt, and most of all Vickie Kearn, senior mathematics editor at Princeton University Press, who oversaw this process with her special combination of expertise and friendship.

Lastly, I thank my wife and colleague Penny Dunham. She created the book's diagrams and provided helpful suggestions as to its contents. Her presence has made this understanding, and the past 35 years, so much fun.

W. Dunham
Allentown, PA

THE CALCULUS
GALLERY

INTRODUCTION

"The calculus," wrote John von Neumann (1903–1957), "was the first achievement of modern mathematics, and it is difficult to overestimate its importance" [1].

Today, more than three centuries after its appearance, calculus continues to warrant such praise. It is the bridge that carries students from the basics of elementary mathematics to the challenges of higher mathematics and, as such, provides a dazzling transition from the finite to the infinite, from the discrete to the continuous, from the superficial to the profound. So esteemed is calculus that its name is often preceded by "the," as in von Neumann's observation above. This gives "*the* calculus" a status akin to "*the* law"—that is, a subject vast, self-contained, and awesome.

Like any great intellectual pursuit, the calculus has a rich history and a rich *pre*history. Archimedes of Syracuse (ca. 287–212 BCE) found certain areas, volumes, and surfaces with a technique we now recognize as proto-integration. Much later, Pierre de Fermat (1601–1665) determined slopes of tangents and areas under curves in a remarkably modern fashion. These and many other illustrious predecessors brought calculus to the threshold of existence.

Nevertheless, this book is not about forerunners. It goes without saying that calculus owes much to those who came before, just as modern art owes much to the artists of the past. But a specialized museum—the Museum of Modern Art, for instance—need not devote room after room to premodern influences. Such an institution can, so to speak, start in the middle. And so, I think, can I.

Thus I shall begin with the two seventeenth-century scholars, Isaac Newton (1642–1727) and Gottfried Wilhelm Leibniz (1646–1716), who gave birth to the calculus. The latter was first to publish his work in a 1684 paper whose title contained the Latin word *calculi* (a system of calculation) that would attach itself to this new branch of mathematics. The first textbook appeared a dozen years later, and the calculus was here to stay.

As the decades passed, others took up the challenge. Prominent among these pioneers were the Bernoulli brothers, Jakob (1654–1705) and Johann (1667–1748), and the incomparable Leonhard Euler (1707–1783), whose research filled many thousands of pages with mathematics of the highest quality. Topics under consideration expanded to include limits, derivatives, integrals, infinite sequences, infinite series, and more. This extended body of material has come to be known under the general rubric of "analysis."

With increased sophistication came troubling questions about the underlying logic. Despite the power and utility of calculus, it rested upon a less-than-certain foundation, and mathematicians recognized the need to recast the subject in a precise, rigorous fashion after the model of Euclid's geometry. Such needs were addressed by nineteenth-century analysts like Augustin-Louis Cauchy (1789–1857), Georg Friedrich Bernhard Riemann (1826–1866), Joseph Liouville (1809–1882), and Karl Weierstrass (1815–1897). These individuals worked with unprecedented care, taking pains to define their terms exactly and to prove results that had hitherto been accepted uncritically.

But, as often happens in science, the resolution of one problem opened the door to others. Over the last half of the nineteenth century, mathematicians employed these logically rigorous tools in concocting a host of strange counterexamples, the understanding of which pushed analysis ever further toward generality and abstraction. This trend was evident in the set theory of Georg Cantor (1845–1918) and in the subsequent achievements of scholars like Vito Volterra (1860–1940), René Baire (1874–1932), and Henri Lebesgue (1875–1941).

By the early twentieth century, analysis had grown into an enormous collection of ideas, definitions, theorems, and examples—and had developed a characteristic manner of thinking—that established it as a mathematical enterprise of the highest rank.

What follows is a sampler from that collection. My goal is to examine the handiwork of those individuals mentioned above and to do so in a manner faithful to the originals yet comprehensible to a modern reader. I shall discuss theorems illustrating the development of calculus over its formative years and the genius of its most illustrious practitioners. The book will be, in short, a "great theorems" approach to this fascinating story.

To this end I have restricted myself to the work of a few representative mathematicians. At the outset I make a full disclosure: my cast of characters was dictated by personal taste. Some whom I have included, like Newton, Cauchy, Weierstrass, would appear in any book with similar objectives. Some, like Liouville, Volterra and Baire, are more idiosyncratic. And others, like Gauss, Bolzano, and Abel, failed to make my cut.

Likewise, some of the theorems I discuss are known to any mathematically literate reader, although their *original* proofs may come as a surprise to those not conversant with the history of mathematics. Into this category fall Leibniz's barely recognizable derivation of the "Leibniz series" from 1673 and Cantor's first but less-well-known proof of the nondenumerability of the continuum from 1874. Other theorems, although part of the folklore of mathematics, seldom appear in modern textbooks; here I am thinking of a result like Weierstrass's everywhere continuous, nowhere differentiable function that so astounded the mathematical world when it was presented to the Berlin Academy in 1872. And some of my choices, I concede, are downright quirky. Euler's evaluation of $\int_0^1 \frac{\sin(\ln x)}{\ln x} dx$, for example, is included simply as a demonstration of his analytic wizardry.

Each result, from Newton's derivation of the sine series to the appearance of the gamma function to the Baire category theorem, stood at the research frontier of its day. Collectively, they document the evolution of analysis over time, with the attendant changes in style and substance. This evolution is striking, for the difference between a theorem from Lebesgue in 1904 and one from Leibniz in 1690 can be likened to the difference between modern literature and *Beowulf*. Nonetheless—and this is critical— I believe that each theorem reveals an ingenuity worthy of our attention and, even more, of our admiration.

Of course, trying to characterize analysis by examining a few theorems is like trying to characterize a thunderstorm by collecting a few raindrops. The impression conveyed will be hopelessly incomplete. To undertake such a project, an author must adopt some fairly restrictive guidelines.

One of mine was to resist writing a comprehensive history of analysis. That is far too broad a mission, and, in any case, there are many works that describe the development of calculus. Some of my favorites are mentioned explicitly in the text or appear as sources in the notes at the end of the book.

A second decision was to exclude topics from both multivariate calculus and complex analysis. This may be a regrettable choice, but I believe it is a defensible one. It has imposed some manageable boundaries upon the contents of the book and thereby has added coherence to the tale. Simultaneously, this restriction should minimize demands upon the reader's background, for a volume limited to topics from *univariate, real* analysis should be understandable to the widest possible audience.

This raises the issue of prerequisites. The book's objectives dictate that I include much technical detail, so the mathematics necessary to follow these theorems is substantial. Some of the early results require consider-able algebraic stamina in chasing formulas across the page. Some of the later ones demand a refined sense of abstraction. All in all, I would not recommend this for the mathematically faint-hearted.

At the same time, in an attempt to favor clarity over conciseness, I have adopted a more conversational style than one would find in a stan-dard text. I intend that the book be accessible to those who have majored or minored in college mathematics and who are not put off by an integral here or an epsilon there. My goal is to keep the prerequisites as modest as the topics permit, but no less so. To do otherwise, to water down the con-tent, would defeat my broader purpose.

So, this is not primarily a biography of mathematicians, nor a history of calculus, nor a textbook. I say this despite the fact that at times I pro-vide biographical information, at times I discuss the history that ties one topic to another, and at times I introduce unfamiliar (or perhaps long for-gotten) ideas in a manner reminiscent of a textbook. But my foremost motivation is simple: to share some favorite results from the rich history of analysis.

And this brings me to a final observation.

In most disciplines there is a tradition of studying the major works of illustrious predecessors, the so-called "masters" of the field. Students of lit-erature read Shakespeare; students of music listen to Bach. In mathematics such a tradition is, if not entirely absent, at least fairly uncommon. This book is meant to address that situation. Although it is not intended as a *his-tory* of the calculus, I have come to regard it as a *gallery* of the calculus.

To this end, I have assembled a number of masterpieces, although these are not the paintings of Rembrandt or Van Gogh but the theorems of Euler or Riemann. Such a gallery may be a bit unusual, but its objective is that of all worthy museums: to serve as a repository of excellence.

Like any gallery, this one has gaps in its collection. Like any gallery, there is not space enough to display all that one might wish. These limi-tations notwithstanding, a visitor should come away enriched by an appreciation of genius. And, in the final analysis, those who stroll among the exhibits should experience the mathematical imagination at its most profound.

♕

CHAPTER I

Newton

Isaac Newton

Isaac Newton (1642–1727) stands as a seminal figure not just in mathematics but in all of Western intellectual history. He was born into a world where science had yet to establish a clear supremacy over medieval superstition. By the time of his death, the Age of Reason was in full bloom. This remarkable transition was due in no small part to his own contributions.

For mathematicians, Isaac Newton is revered as the creator of calculus, or, to use his name for it, of "fluxions." Its origin dates to the mid-1660s when he was a young scholar at Trinity College, Cambridge. There he had absorbed the work of such predecessors as René Descartes (1596–1650), John Wallis (1616–1703), and Trinity's own Isaac Barrow (1630–1677), but he soon found himself moving into uncharted territory. During the next few years, a period his biographer Richard Westfall characterized as one of "incandescent activity," Newton changed forever the mathematical landscape [1]. By 1669, Barrow himself was describing his colleague as "a fellow of our College and very young . . . but of an extraordinary genius and proficiency" [2].

In this chapter, we look at a few of Newton's early achievements: his generalized binomial expansion for turning certain expressions into infinite series, his technique for finding inverses of such series, and his quadrature rule for determining areas under curves. We conclude with a spectacular consequence of these: the series expansion for the sine of an angle. Newton's account of the binomial expansion appears in his *epistola prior*, a letter he sent to Leibniz in the summer of 1676 long after he had done the original work. The other discussions come from Newton's 1669 treatise *De analysi per aequationes numero terminorum infinitas*, usually called simply the *De analysi*.

Although this chapter is restricted to Newton's early work, we note that "early" Newton tends to surpass the mature work of just about anyone else.

GENERALIZED BINOMIAL EXPANSION

By 1665, Isaac Newton had found a simple way to expand—his word was "reduce"—binomial expressions into series. For him, such reductions would be a means of recasting binomials in alternate form as well as an entryway into the method of fluxions. This theorem was the starting point for much of Newton's mathematical innovation.

As described in the *epistola prior*, the issue at hand was to reduce the binomial $(P + PQ)^{m/n}$ and to do so whether m/n "is integral or (so to speak) fractional, whether positive or negative" [3]. This in itself was a bold idea for a time when exponents were sufficiently unfamiliar that they had first to be explained, as Newton did by stressing that "instead of $\sqrt{a}, \sqrt[3]{a}, \sqrt[3]{a^5}$, etc. I write $a^{1/2}, a^{1/3}, a^{5/3}$, and instead of $1/a, 1/aa, 1/a^3$, I write a^{-1}, a^{-2}, a^{-3}" [4]. Apparently readers of the day needed a gentle reminder.

Newton discovered a pattern for expanding not only elementary binomials like $(1 + x)^5$ but more sophisticated ones like $\dfrac{1}{\sqrt[3]{(1 + x)^5}} = (1 + x)^{-5/3}$. The reduction, as Newton explained to Leibniz, obeyed the rule

$$(P + PQ)^{m/n} = P^{m/n} + \frac{m}{n} AQ + \frac{m - n}{2n} BQ$$
$$+ \frac{m - 2n}{3n} CQ + \frac{m - 3n}{4n} DQ + \text{etc.}, \tag{1}$$

where each of A, B, C, . . . represents the previous term, as will be illustrated below. This is his famous binomial expansion, although perhaps in an unfamiliar guise.

Newton provided the example of $\sqrt{c^2 + x^2} = [c^2 + c^2(x^2/c^2)]^{1/2}$. Here, $P = c^2$, $Q = \dfrac{x^2}{c^2}$, $m = 1$, and $n = 2$. Thus,

$$\sqrt{c^2 + x^2} = (c^2)^{1/2} + \frac{1}{2}A\frac{x^2}{c^2} - \frac{1}{4}B\frac{x^2}{c^2} - \frac{1}{2}C\frac{x^2}{c^2}$$
$$- \frac{5}{8}D\frac{x^2}{c^2} - \cdots.$$

To identify A, B, C, and the rest, we recall that each is the immediately preceding term. Thus, $A = (c^2)^{1/2} = c$, giving us

$$\sqrt{c^2 + x^2} = c + \frac{x^2}{2c} - \frac{1}{4}B\frac{x^2}{c^2} - \frac{1}{2}C\frac{x^2}{c^2} - \frac{5}{8}D\frac{x^2}{c^2} - \cdots.$$

Likewise B is the previous term—i.e., $B = \dfrac{x^2}{2c}$—so at this stage we have

$$\sqrt{c^2 + x^2} = c + \frac{x^2}{2c} - \frac{x^4}{8c^3} - \frac{1}{2}C\frac{x^2}{c^2} - \frac{5}{8}D\frac{x^2}{c^2} - \cdots.$$

The analogous substitutions yield $C = -\dfrac{x^4}{8c^3}$ and then $D = \dfrac{x^6}{16c^5}$. Working from left to right in this fashion, Newton arrived at

$$\sqrt{c^2 + x^2} = c + \frac{x^2}{2c} - \frac{x^4}{8c^3} + \frac{x^6}{16c^5} - \frac{5x^8}{128c^7} + \cdots.$$

Obviously, the technique has a recursive flavor: one finds the coefficient of x^8 from the coefficient of x^6, which in turn requires the coefficient of x^4, and so on. Although the modern reader is probably accustomed to a "direct" statement of the binomial theorem, Newton's recursion has an undeniable appeal, for it streamlines the arithmetic when calculating a numerical coefficient from its predecessor.

For the record, it is a simple matter to replace A, B, C, . . . by their equivalent expressions in terms of P and Q, then factor the common

$P^{m/n}$ from both sides of (1), and so arrive at the result found in today's texts:

$$(1+Q)^{m/n} = 1 + \frac{m}{n}Q + \frac{\frac{m}{n}\left(\frac{m}{n}-1\right)}{2\times 1}Q^2$$

$$+ \frac{\frac{m}{n}\left(\frac{m}{n}-1\right)\left(\frac{m}{n}-2\right)}{3\times 2\times 1}Q^3 + \cdots . \tag{2}$$

Newton likened such reductions to the conversion of square roots into infinite decimals, and he was not shy in touting the benefits of the operation. "It is a convenience attending infinite series," he wrote in 1671,

> that all kinds of complicated terms . . . may be reduced to the class of simple quantities, i.e., to an infinite series of fractions whose numerators and denominators are simple terms, which will thus be freed from those difficulties that in their original form seem'd almost insuperable. [5]

To be sure, freeing mathematics from insuperable difficulties is a worthy undertaking.

One additional example may be helpful. Consider the expansion of $\frac{1}{\sqrt{1-x^2}}$, which Newton put to good use in a result we shall discuss later in the chapter. We first write this as $(1-x^2)^{-1/2}$, identify $m=-1$, $n=2$, and $Q=-x^2$, and apply (2):

$$\frac{1}{\sqrt{1-x^2}} = 1 + \left(-\frac{1}{2}\right)(-x^2) + \frac{(-1/2)(-3/2)}{2\times 1}(-x^2)^2$$

$$+ \frac{(-1/2)(-3/2)(-5/2)}{3\times 2\times 1}(-x^2)^3$$

$$+ \frac{(-1/2)(-3/2)(-5/2)(-7/2)}{4\times 3\times 2\times 1}(-x^2)^4 + \cdots$$

$$= 1 + \frac{1}{2}x^2 + \frac{3}{8}x^4 + \frac{5}{16}x^6 + \frac{35}{128}x^8 + \cdots . \tag{3}$$

Newton would "check" an expansion like (3) by *squaring* the series and examining the answer. If we do the same, restricting our attention to terms of degree no higher than x^8, we get

$$\left[1 + \frac{1}{2}x^2 + \frac{3}{8}x^4 + \frac{5}{16}x^6 + \frac{35}{128}x^8 + \cdots\right]$$

$$\times \left[1 + \frac{1}{2}x^2 + \frac{3}{8}x^4 + \frac{5}{16}x^6 + \frac{35}{128}x^8 + \cdots\right]$$

$$= 1 + x^2 + x^4 + x^6 + x^8 + \cdots,$$

where all of the coefficients miraculously turn out to be 1 (try it!). The resulting product, of course, is an infinite geometric series with common ratio x^2 which, by the well-known formula, sums to $\dfrac{1}{1-x^2}$. But if the *square* of the series in (3) is $\dfrac{1}{1-x^2}$, we conclude that that series itself must be $\dfrac{1}{\sqrt{1-x^2}}$. *Voila!*

Newton regarded such calculations as compelling evidence for his general result. He asserted that the "common analysis performed by means of equations of a finite number of terms" may be extended to such infinite expressions "albeit we mortals whose reasoning powers are confined within narrow limits, can neither express nor so conceive all the terms of these equations, as to know exactly from thence the quantities we want" [6].

INVERTING SERIES

Having described a method for reducing certain binomials to infinite series of the form $z = A + Bx + Cx^2 + Dx^3 + \cdots$, Newton next sought a way of finding the series for x in terms of z. In modern terminology, he was seeking the inverse relationship. The resulting technique involves a bit of heavy algebraic lifting, but it warrants our attention for it too will appear later on. As Newton did, we describe the inversion procedure by means of a specific example.

Beginning with the series $z = x - x^2 + x^3 - x^4 + \cdots$, we rewrite it as

$$(x - x^2 + x^3 - x^4 + \cdots) - z = 0 \tag{4}$$

and discard all powers of x greater than or equal to the quadratic. This, of course, leaves $x - z = 0$, and so the inverted series begins as $x = z$.

Newton was aware that discarding all those higher degree terms rendered the solution inexact. The *exact* answer would have the form $x = z + p$, where p is a series yet to be determined. Substituting $z + p$ for x in (4) gives

$$[(z + p) - (z + p)^2 + (z + p)^3 - (z + p)^4 + \cdots] - z = 0,$$

which we then expand and rearrange to get

$$[-z^2 + z^3 - z^4 + z^5 - \cdots] + [1 - 2z + 3z^2 - 4z^3 + 5z^4 - \cdots]p$$
$$+ [-1 + 3z - 6z^2 + 10z^3 - \cdots]p^2 + [1 - 4z + 10z^2 - \cdots]p^3$$
$$+ [-1 + 5z - \cdots]p^4 + \cdots = 0. \tag{5}$$

Next, jettison the quadratic, cubic, and higher degree terms in p and solve to get

$$p = \frac{z^2 - z^3 + z^4 - z^5 + \cdots}{1 - 2z + 3z^2 - 4z^3 + \cdots}.$$

Newton now did a second round of weeding, as he tossed out all but the lowest power of z in numerator and denominator. Hence p is approximately $\dfrac{z^2}{1}$, so the inverted series at this stage looks like $x = z + p = z + z^2$.

But p is not *exactly* z^2. Rather, we say $p = z^2 + q$, where q is a series to be determined. To do so, we substitute into (5) to get

$$[-z^2 + z^3 - z^4 + z^5 - \cdots] + [1 - 2z + 3z^2 - 4z^3 + 5z^4 - \cdots](z^2 + q)$$
$$+ [-1 + 3z - 6z^2 + 10z^3 - \cdots](z^2 + q)^2 + [1 - 4z + 10z^2 - \cdots]$$
$$(z^2 + q)^3 + [-1 + 5z - \cdots](z^2 + q)^4 + \cdots = 0.$$

We expand and collect terms by powers of q:

$$[-z^3 + z^4 - z^6 + \cdots] + [1 - 2z + z^2 + 2z^3 - \cdots]q$$
$$+ [-1 + 3z - 3z^2 - 2z^3 + \cdots]q^2 + \cdots. \tag{6}$$

As before, discard terms involving powers of q above the first, solve to get $q = \dfrac{z^3 - z^4 + z^6 - \cdots}{1 - 2z + z^2 + 2z^3 + \cdots}$, and then drop all but the lowest degree terms top and bottom to arrive at $q = \dfrac{z^3}{1}$. At this point, the series looks like $x = z + z^2 + q = z + z^2 + z^3$.

The process would be continued by substituting $q = z^3 + r$ into (6). Newton, who had a remarkable tolerance for algebraic monotony, seemed able to continue such calculations *ad infinitum* (almost). But eventually even he was ready to step back, examine the output, and seek a pattern. Newton put it this way: "Let it be observed here, by the bye, that when 5 or 6 terms . . . are known, they may be continued at pleasure for most part, by observing the analogy of the progression" [7].

For our example, such an examination suggests that $x = z + z^2 + z^3 + z^4 + z^5 + \cdots$ is the inverse of the series $z = x - x^2 + x^3 - x^4 + \cdots$ with which we began.

In what sense can this be trusted? After all, Newton discarded most of his terms most of the time, so what confidence remains that the answer is correct?

Again, we take comfort in the following "check." The original series $z = x - x^2 + x^3 - x^4 + \cdots$ is geometric with common ratio $-x$, and so in closed form $z = \dfrac{x}{1+x}$. Consequently, $x = \dfrac{z}{1-z}$, which we recognize to be the sum of the geometric series $z + z^2 + z^3 + z^4 + z^5 + \cdots$. This is precisely the result to which Newton's procedure had led us. Everything seems to be in working order.

The techniques encountered thus far—the generalized binomial expansion and the inversion of series—would be powerful tools in Newton's hands. There remains one last prerequisite, however, before we can truly appreciate the master at work.

QUADRATURE RULES FROM THE *DE ANALYSI*

In his *De analysi* of 1669, Newton promised to describe the method "which I had devised some considerable time ago, for measuring the quantity of curves, by means of series, infinite in the number of terms" [8]. This was not Newton's first account of his fluxional discoveries, for he had drafted an October 1666 tract along these same lines. The *De analysi* was a revision that displayed the polish of a maturing thinker. Modern scholars find it strange that the secretive Newton withheld this manuscript from all but a few lucky colleagues, and it did not appear in print until 1711, long after many of its results had been published by others. Nonetheless, the early date and illustrious authorship justify its description as "perhaps the most celebrated of all Newton's mathematical writings" [9].

The treatise began with a statement of the three rules for "the quadrature of simple curves." In the seventeenth century, *quadrature* meant determination of area, so these are just integration rules.

> Rule 1. The quadrature of simple curves: If $y = ax^{m/n}$ is the curve
> AD, where a is a constant and m and n are positive integers, then
> the area of region ABD is $\dfrac{an}{m+n} x^{(m+n)/n}$ (see figure 1.1).

A modern version of this would identify A as the origin, B as $(x, 0)$, and the curve as $y = at^{m/n}$. Newton's statement then becomes $\int_0^x at^{m/n} dt = \dfrac{ax^{(m/n)+1}}{(m/n)+1} = \dfrac{an}{m+n} x^{(m+n)/n}$, which is just a special case of the power rule from integral calculus.

Only at the end of the *De analysi* did Newton observe, almost as an afterthought, that "an attentive reader" would want to see a proof for Rule 1 [10]. Attentive as always, we present his argument below.

Again, let the curve be AD with $AB = x$ and $BD = y$, as shown in figure 1.2. Newton assumed that the *area ABD* beneath the curve was given by an expression z written in terms of x. The goal was to find a corresponding

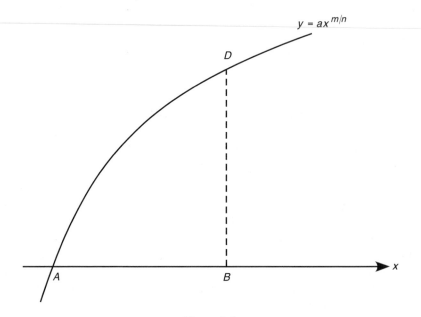

Figure 1.1

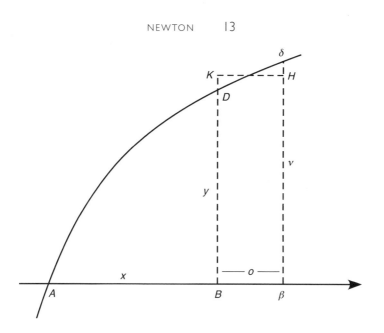

Figure 1.2

formula for y in terms of x. From a modern vantage point, he was beginning with $z = \int_0^x y(t)dt$ and seeking $y = y(x)$. His derivation blended geometry, algebra, and fluxions before ending with a few dramatic flourishes.

At the outset, Newton let β be a point on the horizontal axis a tiny distance o from B. Thus, segment $A\beta$ has length $x + o$. He let z be the area ABD, although to emphasize the functional relationship we shall take the liberty of writing $z = z(x)$. Hence, $z(x + o)$ is the area $A\beta\delta$ under the curve. Next he introduced rectangle $B\beta HK$ of height $v = BK = \beta H$, the area of which he stipulated to be *exactly* that of region $B\beta\delta D$ beneath the curve. In other words, the area of $B\beta\delta D$ was to be ov.

At this point, Newton specified that $z(x) = \dfrac{an}{m+n}x^{(m+n)/n}$ and proceeded to find the instantaneous rate of change of z. To do so, he examined the change in z divided by the change in x as the latter becomes small. For notational ease, he temporarily let $c = an/(m + n)$ and $p = m + n$ so that $z(x) = cx^{p/n}$ and

$$[z(x)]^n = c^n x^p. \tag{7}$$

Now, $z(x + o)$ is the area $A\beta\delta$, which can be decomposed into the area of ABD and that of $B\beta\delta D$. The latter, as noted, is the same as rectangular area ov and so Newton concluded that $z(x + o) = z(x) + ov$. Substituting into (7), he got

$$[z(x) + ov]^n = [z(x + o)]^n = c^n(x + o)^p,$$

and the binomials on the left and right were expanded to

$$[z(x)]^n + n[z(x)]^{n-1}ov + \frac{n(n-1)}{2}[z(x)]^{n-2}o^2v^2 + \cdots$$

$$= c^n x^p + c^n px^{p-1}o + c^n \frac{p(p-1)}{2}x^{p-2}o^2 + \cdots.$$

Applying (7) to cancel the leftmost terms on each side and then dividing through by o, Newton arrived at

$$n[z(x)]^{n-1}v + \frac{n(n-1)}{2}[z(x)]^{n-2}ov^2 + \cdots$$

$$= c^n px^{p-1} + c^n \frac{p(p-1)}{2}x^{p-2}o + \cdots. \tag{8}$$

At that point, he wrote, "If we suppose $B\beta$ to be diminished infinitely and to vanish, or o to be nothing, v and y in that case will be equal, and the terms which are multiplied by o will vanish" [11]. He was asserting that, as o becomes zero, so do all terms in (8) that contain o. At the same time, v becomes equal to y, which is to say that the height BK of the rectangle in Figure 1.2 will equal the ordinate BD of the original curve. In this way, (8) transforms into

$$n[z(x)]^{n-1}y = c^n px^{p-1}. \tag{9}$$

A modern reader is likely to respond, "Not so fast, Isaac!" When Newton divided by o, that quantity most certainly was *not* zero. A moment later, it *was* zero. There, in a nutshell, lay the rub. This zero/nonzero dichotomy would trouble analysts for the next century and then some. We shall have much more to say about this later in the book.

But Newton proceeded. In (9) he substituted for $z(x)$, c, and p and solved for

$$y = \frac{c^n px^{p-1}}{n[z(x)]^{n-1}} = \frac{\left[\dfrac{an}{(m+n)}\right]^n (m+n)x^{m+n-1}}{n\left[\dfrac{an}{(m+n)}x^{(m+n)/n}\right]^{n-1}} = ax^{m/n}.$$

Thus, starting from his assumption that the area *ABD* is given by $z(x) = \dfrac{an}{m+n} x^{(m+n)/n}$. Newton had deduced that curve *AD* must satisfy the equation $y = ax^{m/n}$. He had, in essence, differentiated the integral. Then, without further justification, he stated, "Wherefore conversely, if $ax^{m/n} = y$, it shall be $\dfrac{an}{m+n} x^{(m+n)/n} = z$." His proof of rule 1 was finished [12].

This was a peculiar twist of logic. Having derived the equation of y from that of its area accumulator z, Newton asserted that the relationship went the other way and that the area under $y = ax^{m/n}$ is indeed $\dfrac{an}{m+n} x^{(m+n)/n}$. Such an argument tends to leave us with mixed feelings, for it features some gaping logical chasms. Derek Whiteside, editor of Newton's mathematical papers, aptly characterized this quadrature proof as "a brief, scarcely comprehensible appearance of fluxions" [13]. On the other hand, it is important to remember the source. Newton was writing at the very beginning of the long calculus journey. Within the context of his time, the proof was groundbreaking, and his conclusion was correct. Something rings true in Richard Westfall's observation that, "however briefly, *De analysi* did indicate the full extent and power of the fluxional method" [14].

Whatever the modern verdict, Newton was satisfied. His other two rules, for which the *De analysi* contained no proofs, were as follows:

> Rule 2. The quadrature of curves compounded of simple ones: If the value of y be made up of several such terms, the area likewise shall be made up of the areas which result from every one of the terms. [15]

> Rule 3. The quadrature of all other curves: But if the value of y, or any of its terms be more compounded than the foregoing, it must be reduced into more simple terms . . . and afterwards by the preceding rules you will discover the [area] of the curve sought. [16]

Newton's second rule affirmed that the integral of the sum of finitely many terms is the sum of the integrals. This he illustrated with an example or two. The third rule asserted that, when confronted with a more complicated expression, one was first to "reduce" it into an infinite series, integrate each term of the series by means of the first rule, and then sum the results.

This last was an appealing idea. More to the point, it was the final prerequisite Newton would need to derive a mathematical blockbuster: the infinite series for the sine of an angle. This great theorem from the *De analysi* will serve as the chapter's climax.

Newton's Derivation of the Sine Series

Consider in figure 1.3 the quadrant of a circle centered at the origin and with radius 1, where as before $AB = x$ and $BD = y$. Newton's initial objective was to find an expression for the length of arc αD [17].

From D, draw DT tangent to the circle, and let BK be "the moment of the base AB." In a notation that would become standard after Newton, we let $BK = dx$. This created the "indefinitely small" right triangle DGH, whose hypotenuse DH Newton regarded as the moment of the arc αD. We write $DH = dz$, where $z = z(x)$ stands for the length of arc αD. Because all of this is occurring within the unit circle, the radian measure of $\angle \alpha AD$ is z as well.

Under this scenario, the infinitely small triangle DGH is similar to triangle DBT so that $\dfrac{GH}{DH} = \dfrac{BT}{DT}$. Moreover, radius AD is perpendicular to tangent line DT, and so altitude BD splits right triangle ADT into similar pieces: triangles DBT and ABD. It follows that $\dfrac{BT}{DT} = \dfrac{BD}{AD}$, and from these two proportions we conclude that $\dfrac{GH}{DH} = \dfrac{BD}{AD}$. With the differential notation above, this amounts to $\dfrac{dx}{dz} = \dfrac{y}{1}$, and hence $dz = \dfrac{dx}{y}$.

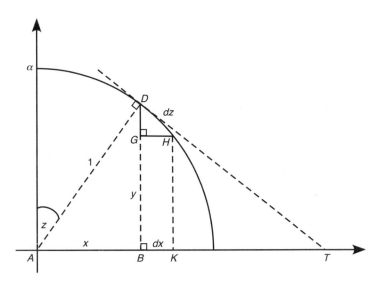

Figure 1.3

Newton's next step was to exploit the circular relationship $y = \sqrt{1 - x^2}$ to conclude that $dz = \dfrac{dx}{y} = \dfrac{dx}{\sqrt{1 - x^2}}$. Expanding $\dfrac{1}{\sqrt{1 - x^2}}$ as in (3) led to

$$dz = \left[1 + \frac{1}{2}x^2 + \frac{3}{8}x^4 + \frac{5}{16}x^6 + \frac{35}{128}x^8 + \cdots \right]dx,$$

and so

$$z = z(x) = \int_0^x dz = \int_0^x \left[1 + \frac{1}{2}t^2 + \frac{3}{8}t^4 + \frac{5}{16}t^6 + \frac{35}{128}t^8 + \cdots \right]dt.$$

Finding the quadratures of these individual powers and summing the results by Rule 3, Newton concluded that the arclength of αD was

$$z = x + \frac{1}{6}x^3 + \frac{3}{40}x^5 + \frac{5}{112}x^7 + \frac{35}{1152}x^9 + \cdots. \tag{10}$$

Referring again to Figure 1.3, we see that z is not only the radian measure of $\angle \alpha AD$, but the measure of $\angle ADB$ as well. From triangle ABD, we know that $\sin z = x$ and so

$$\arcsin x = z = x + \frac{1}{6}x^3 + \frac{3}{40}x^5 + \frac{5}{112}x^7 + \frac{35}{1152}x^9 + \cdots.$$

Thus, beginning with the *algebraic* expression $\dfrac{1}{\sqrt{1 - x^2}}$, Newton had used his generalized binomial expansion and basic integration to derive the series for arcsine, an intrinsically more complicated relationship.

But Newton had one other trick up his sleeve. Instead of a series for arclength (z) in terms of its coordinate (x), he sought to reverse the process. He wrote, "If, from the Arch αD given, the Sine AB was required, I extract the root of the equation found above" [18]. That is, Newton would apply his inversion procedure to convert the series for $z = \arcsin x$ into one for $x = \sin z$.

Following the technique described earlier, we begin with $x = z$ as the first term. To push the expansion to the next step, substitute $x = z + p$ into (10) and solve to get

$$p = \frac{-\dfrac{1}{6}z^3 - \dfrac{3}{40}z^5 - \dfrac{5}{112}z^7 - \cdots}{1 + \dfrac{1}{2}z^2 + \dfrac{3}{8}z^4 + \dfrac{5}{16}z^6 + \cdots},$$

from which we retain only $p = -\dfrac{1}{6}z^3$. This extends the series to $x = z - \dfrac{1}{6}z^3$. Next introduce $p = -\dfrac{1}{6}z^3 + q$ and continue the inversion process, solving for

$$q = \frac{\dfrac{1}{120}z^5 + \dfrac{1}{56}z^7 - \dfrac{1}{72}z^8 + \cdots}{1 + \dfrac{1}{2}z^2 + \dfrac{3}{8}z^4 + \cdots},$$

or simply $q = \dfrac{1}{120}z^5$. At this stage $x = z - \dfrac{1}{6}z^3 + \dfrac{1}{120}z^5$, and, as Newton might say, we "continue at pleasure" until discerning the pattern and writing down one of the most important series in analysis:

$$\sin z = z - \frac{1}{6}z^3 + \frac{1}{120}z^5 - \frac{1}{5040}z^7 + \frac{1}{362880}z^9 - \cdots$$

$$= \sum_{k=0}^{\infty} \frac{(-1)^k}{(2k+1)!}z^{2k+1}.$$

To find the Base from the Length of the Curve given.

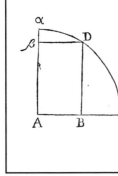

45. If from the Arch αD given the Sine AB was required; I extract the Root of the Equation found above, viz. $z = x + \frac{1}{6}x^3 + \frac{3}{40}x^5 + \frac{5}{112}x^7$ (it being suppofed that AB $= x$, αD $= z$, and A$\alpha = 1$) by which I find $x = z - \frac{1}{6}z^3 + \frac{1}{120}z^5 - \frac{1}{5040}z^7 + \frac{1}{362880}z^9$ &c.

46. And moreover if the Cofine Aβ were required from that Arch given, make Aβ ($= \sqrt{1-xx}$) $= 1 - \frac{1}{2}z^2 + \frac{1}{24}z^4 - \frac{1}{720}z^6 + \frac{1}{40320}z^8 - \frac{1}{3628800}z^{10}$, &c.

Newton's series for sine and cosine (1669)

For good measure, Newton included the series for $\cos z = \displaystyle\sum_{k=0}^{\infty} \frac{(-1)^k}{(2k)!}z^{2k}$. In the words of Derek Whiteside, "These series for the sine and cosine . . . here appear for the first time in a European manuscript" [19].

To us, this development seems incredibly roundabout. We now regard the sine series as a trivial consequence of Taylor's formula and differential calculus. It is so natural a procedure that we expect it was always so. But Newton, as we have seen, approached this very differently. He applied rules of integration, not of differentiation; he generated the sine series from the (to our minds) incidental series for the arcsine; and he needed his complicated inversion scheme to make it all work.

This episode reminds us that mathematics did not necessarily evolve in the manner of today's textbooks. Rather, it developed by fits and starts and odd surprises. Actually that is half the fun, for history is most intriguing when it is at once significant, beautiful, and *unexpected*.

On the subject of the unexpected, we add a word about Whiteside's qualification in the passage above. It seems that Newton was not the first to discover a series for the sine. In 1545, the Indian mathematician Nilakantha (1445–1545) described this series and credited it to his even more remote predecessor Madhava, who lived around 1400. An account of these discoveries, and of the great Indian tradition in mathematics, can be found in [20] and [21]. It is certain, however, that these results were unknown in Europe when Newton was active.

We end with two observations. First, Newton's *De analysi* is a true classic of mathematics, belonging on the bookshelf of anyone interested in how calculus came to be. It provides a glimpse of one of history's most fertile thinkers at an early stage of his intellectual development.

Second, as should be evident by now, a revolution had begun. The young Newton, with a skill and insight beyond his years, had combined infinite series and fluxional methods to push the frontiers of mathematics in new directions. It was his contemporary, James Gregory (1638–1675), who observed that the elementary methods of the past bore the same relationship to these new techniques "as dawn compares to the bright light of noon" [22]. Gregory's charming description was apt, as we see time and again in the chapters to come. And first to travel down this exciting path was Isaac Newton, truly "a man of extraordinary genius and proficiency."

CHAPTER 2

Leibniz

Gottfried Wilhelm Leibniz

Calculus may be unique in having as its founders two individuals better known for other things. In the public mind, Isaac Newton tends to be regarded as a physicist, and his cocreator, Gottfried Wilhelm Leibniz (1646–1716), is likely to be thought of as a philosopher. This is both annoying and flattering—annoying in its disregard for their mathematical contributions and flattering in its recognition that it took more than just an *ordinary* genius to launch the calculus.

Leibniz, with his varied interests and far-reaching contributions, had an intellect of phenomenal breadth. Besides philosophy and mathematics, he excelled in history, jurisprudence, languages, theology, logic, and diplomacy. When only 27, he was admitted to London's Royal Society for inventing a mechanical calculator that added, subtracted, multiplied, and divided—a machine that was by all accounts as revolutionary as it was complicated [1].

Like Newton, Leibniz had an intense period of mathematical activity, although his came later than Newton's and in a different country. Whereas Newton developed his fluxional ideas at Cambridge in the mid-1660s, Leibniz did his groundbreaking work while on a diplomatic mission to Paris a decade later. This gave Newton temporal priority—which he and his countrymen would later assert was the only kind that mattered—but it was Leibniz who published his calculus at a time when the *De analysi* and other Newtonian treatises were gathering dust in manuscript form. Much has been written about the ensuing dispute over which of the two deserved credit for the calculus, and the story is not a pretty one [2]. Modern scholars, centuries removed from passions both national and personal, recognize that the discoveries of Newton and Leibniz were made independently. Like an idea whose time had come, calculus was "in the air" and needed only a remarkably penetrating and integrative mind to bring it into existence. This Newton had.

Just as surely, so did Leibniz. Upon his arrival in Paris in 1672, he was a novice who admitted to lacking "the patience to read through the long series of proofs" necessary for mathematical success [3]. Dissatisfied with his modest knowledge, he spent time filling gaps, reading mathematicians as venerable as Euclid (ca. 300 BCE) or as up-to-date as Pascal (1623–1662), Barrow, and his sometime-mentor, Christiaan Huygens (1629–1695). At first it was hard going, but Leibniz persevered. He recalled that, in spite of his deficiencies, "it seemed to me, I do not know by what rash confidence in my own ability, that I might become the equal of these if I so desired" [4].

Progress was breathtaking. He wrote in one memorable passage that soon he was "ready to get along without help, for I read [mathematics] almost as one reads tales of romance" [5]. After absorbing, almost *inhaling*, the work of his contemporaries, Leibniz pushed beyond them all to create the calculus, thereby earning himself mathematical immortality.

And, unlike Newton across the English Channel, Leibniz was willing to publish. The first printed version of the calculus was Leibniz's 1684 paper bearing the long title, *"Nova methodus pro maximis et minimis, itemque tangentibus, quae nec fractas, nec irrationales quantitates moratur, et singulare pro illis calculi genus."* This translates into "A New Method for Maxima and Minima, and also Tangents, which is Impeded Neither by Fractional Nor by Irrational Quantities, and a Remarkable Type of Calculus for This" [6]. With references to maxima, minima, and tangents, it should come as no surprise that the article was Leibniz's introduction to *differential* calculus. He followed it two years later with a paper on integral calculus. Even at that early stage, Leibniz not only had organized and codified many of the basic calculus rules, but he was already using dx for the differential of x and $\int x\, dx$ for its integral. Among his other talents was his ability to provide what Laplace later called "a very happy notation" [7].

MENSIS OCTOBRIS A. MDCLXXXIV. 467

NOVA METHODVS PRO MAXIMIS ET MI-nimis, itemque tangentibus, quæ nec fractas, nec irrationales quantitates moratur, & singulare pro illis calculi genus, per G.G.L.

SIt axis AX, & curvæ plures, ut VV, WW, YY, ZZ, quarum ordi- TAB. XII.
natæ, ad axem normales, VX, WX, YX, ZX, quæ vocentur respective, *v*, vv, y, z; & ipsa AX abscissa ab axe, vocetur x. Tangentes sint
VB, WC, YD, ZE axi occurrentes respective in punctis B, C, D, E.
Jam recta aliqua pro arbitrio assumta vocetur dx, & recta quæ sit ad
dx, ut *v* (vel vv, vel y, vel z) est ad VB (vel WC, vel YD, vel ZE) vocetur d*v* (vel dvv, vel dy vel dz) sive differentia ipsarum *v* (vel ipsarum vv, aut y, aut z) His positis calculi regulæ erunt tales:

Sit a quantitas data constans, erit da æqualis o, & d̅ ax erit æqu·
a dx: si fit y æqu. *v* (seu ordinata quævis curvæ YY, æqualis cuivis or-
dinatæ respondenti curvæ VV) erit dy æqu. d*v*. Jam *Additio & Sub-
tractio:* si sit z -y + vv + x æqu. *v*, erit d̅z̅-̅y̅ + vv + x seu d*v*, æqu.
dz -dy + dvv + dx. *Multiplicatio*, d x *v* æqu. x d *v* + *v* dx, seu posito
y æqu. x*v*, fiet dy æqu. x d *v* + *v* d x. In arbitrio enim est vel formulam,
ut x*v*, vel compendio pro ea literam, ut y, adhibere.   Notandum & x
& dx eodem modo in hoc calculo tractari, ut y & dy, vel aliam literam
indeterminatam cum sua differentiali.   Notandum etiam nōn dari
semper regressum a differentiali Æquatione, nisi cum quadam cautio-
                                 v   *v*
ne, de quo alibi.   Porro *Divisio*, d—vel (posito z æqu.   ) d z æqu.
                                                          ̅   ̅
+ *v* dy + y d *v*

Leibniz's first paper on differential calculus (1684)

In this chapter, we examine a pair of theorems from the years 1673–1674. Much of our discussion is drawn from Leibniz's monograph *Historia et origo calculi differentialis*, an account of the events surrounding his creation of the calculus [8]. Our first result, more abstract, is known as the transmutation theorem. Although its geometrical convolutions may not appeal to modern tastes, it reveals his mathematical gift and leads to an early version of what we now call integration by parts. The second result, a consequence of the first, is the so-called "Leibniz Series." Like Newton's work, discussed in the previous chapter, this combined series expansions and basic integration techniques to produce an important and fascinating outcome.

THE TRANSMUTATION THEOREM

Finding areas beneath curves was a hot topic in the middle of the seventeenth century, and this is the subject of the Leibniz transmutation theorem. Suppose, in figure 2.1, we seek the area beneath the curve AB. Leibniz imagined this region as being composed of infinitely many "infinitesimal" rectangles, each of width dx and height y, where the latter varies with the shape of AB.

To us today, the nature of Leibniz's dx is unclear. In the seventeenth century, it was seen as a least possible length, an infinitely small magnitude that could not be further subdivided. But how is such a thing possible? Clearly any length, no matter how razor-thin, can be split in half. Leibniz's explanations in this regard were of no help, for even he became

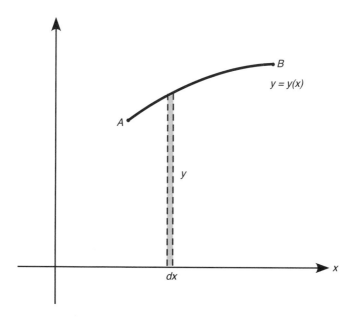

Figure 2.1

unintelligible when addressing the matter. Consider the following passage from sometime after 1684:

> by . . . infinitely small, we understand something . . . indefinitely small, so that each conducts itself as a sort of class, and not merely as the last thing of a class. If anyone wishes to understand these [the infinitely small] as the ultimate things . . . , it can be done, and that too without falling back upon a controversy about the reality of extensions, or of infinite continuums in general, or of the infinitely small, ay even though he think that such things are utterly impossible. [9]

The reader is forgiven for finding this clarification less than clarifying. Leibniz himself seemed to choose expediency over logic when he added that, even if the nature of these indivisibles is uncertain, they can nonetheless be used as "a tool that has advantages for the purpose of the calculation." Again we glimpse the mathematical quagmire that would confront analysts of the future. But in 1673 Leibniz was eager to press on, and a later generation could tidy up the logic.

Returning to figure 2.1, we see that the infinitesimal rectangle has area $y\,dx$. To calculate the area under the curve AB, Leibniz summed an infinitude

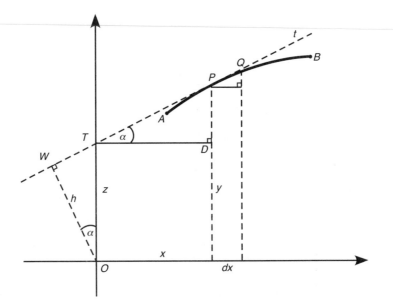

Figure 2.2

of these areas. As a symbol for this process, he chose an elongated "S" (for "*summa*") and thus denoted the area as $\int y\,dx$. Thereafter, his integral sign became the "logo" of calculus, announcing to all who saw it that higher mathematics was afoot.

It is one thing to have a notation for area and quite another to know how to compute it. Leibniz's transmutation theorem was aimed at resolving this latter question.

His idea is illustrated in figure 2.2, which again shows curve AB, the area beneath which is our object. On the curve is an arbitrary point P with coordinates (x, y). At P, Leibniz constructed the tangent line t, meeting the vertical axis at point T with coordinates $(0, z)$. Leibniz explained this construction by noting that "to find a tangent means to draw a line that connects two points of the curve at an infinitely small distance" [10]. Letting dx be an infinitesimal increment in x, he then created an infinitely small right triangle with hypotenuse PQ along the tangent line and having sides of length dx, dy, and ds, an enlargement of which appears in figure 2.3. We let α be the angle of inclination of this tangent line.

Leibniz stressed that, "Even though this triangle is indefinite (being infinitely small), yet . . . it was always possible to find definite triangles similar to it" [11]. Of course, one may wonder how an infinitely small triangle can be similar to *anything*, but this is not the time to quibble. Leibniz regarded ΔTDP in Figure 2.2 as being similar to the infinitesimal triangle in figure 2.3. It followed that $\dfrac{dy}{dx} = \dfrac{PD}{TD} = \dfrac{y - z}{x}$, which he solved to get

$$z = y - x\frac{dy}{dx}. \tag{1}$$

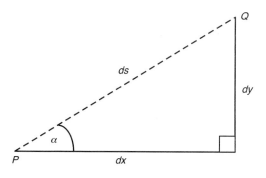

Figure 2.3

Next, Leibniz extended his tangent line *PT* leftward, and from the origin drew segment *OW* of length *h* perpendicular to this extension (again, see figure 2.2). Because ∠*PTD* has measure α, we know that ∠*OTW* has measure $\pi - \alpha$, and so the measure of ∠*TOW* is α as well. This makes Δ*OTW* similar to the infinitesimal triangle, and so we generate another proportion $\dfrac{z}{h} = \dfrac{ds}{dx}$, from which we conclude that

$$h \, ds = z \, dx. \tag{2}$$

Leibniz then drew Δ*OPQ* radiating from the origin and having as base the hypotenuse *PQ* of the infinitesimal triangle. In order not to clutter figure 2.2 any further, we redraw the diagram, with this particular triangle, in figure 2.4.

By now, the reader may suspect that Leibniz was adrift, lost in a sea of pointless triangles. But in fact the oblique, infinitesimal triangle *OPQ* was central to his transmutation theorem. Because its base is of length $\overline{PQ} = ds$ and its height is $\overline{OW} = h$, we see that its area is $\dfrac{1}{2} h \, ds$, which, by (2) above, is just $\dfrac{1}{2} z \, dx$.

Leibniz assembled an infinitude of these infinitesimal triangles, all radiating from the origin and terminating along *AB*, as shown in figure 2.5.

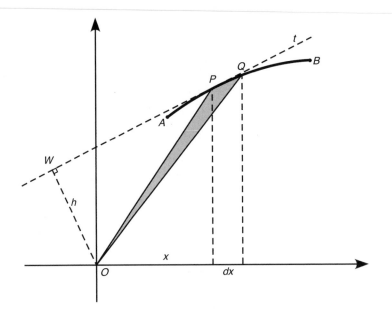

Figure 2.4

Writing years later, Leibniz remembered that he "happened to have occasion to break up an area into triangles formed by a number of straight lines meeting in a point, and . . . perceived that something new could be readily obtained from it" [12].

This polar perspective was critical, for Leibniz recognized that the area of the wedge in figure 2.5 was the sum of the areas of infinitesimal triangles whose analytic expression he had determined above. That is,

$$\text{Area (wedge)} = \text{Sum of triangular areas} = \int \frac{1}{2} z \, dx = \frac{1}{2} \int z \, dx. \quad (3)$$

In truth, Leibniz was not primarily interested in the area of this wedge. Rather, he sought the area under curve AB in figure 2.1, that is, $\int y \, dx$. Fortunately it takes only a bit of tinkering to relate the areas in question, for the geometry of figure 2.6 shows that

Area under curve AB = Area (wedge) + Area (ΔObB) − Area (ΔOaA).

This relationship, by (3), has the symbolic equivalent

$$\int y \, dx = \frac{1}{2} \int z \, dx + \frac{1}{2} b \, y(b) - \frac{1}{2} a \, y(a). \quad (4)$$

Here at last is the transmutation theorem. The name indicates that the original integral $\int y \, dx$ has been transformed (or "transmuted") into a sum

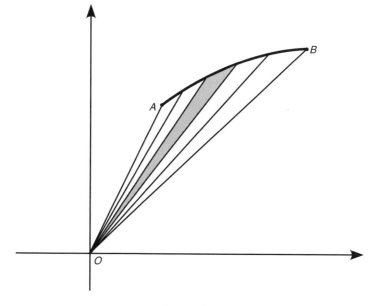

Figure 2.5

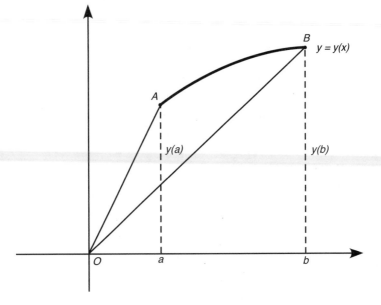

Figure 2.6

of the new integral $\dfrac{1}{2}\int z\,dx$ and the constant $\dfrac{1}{2}b\,y(b)-\dfrac{1}{2}a\,y(a)$. Today we might find it more palatable to insert limits of integration (a notational device Leibniz did not employ) and recast the theorem as

$$\int_a^b y\,dx = \frac{1}{2}\int_a^b z\,dx + \frac{1}{2}\left[xy\Big|_a^b\right].\qquad(5)$$

Formula (5) is notable for at least two reasons.

First, it is possible that the "new" integral in z may be easier to evaluate than the original one in y. If so, z would play an auxiliary role in finding the original area. For seventeenth century mathematicians, a curve playing such a role was called a *quadratrix*, that is, a facilitator of quadrature. If it produced a simpler integral, then this whole, long process would pay off. As we shall see in a moment, this is exactly what happened in the derivation of the Leibniz series.

The relationship in (5) has a theoretical significance as well. Recall that $z = z(x)$ was the y-intercept of the line tangent to the curve AB at the point (x, y). The value of z thus depends on the slope of the tangent line and so injects the derivative into this mix of integrals. One senses that an important connection is lurking in the wings.

To see it, we recall from (1) that $z = y - x\dfrac{dy}{dx}$ and so $z\,dx = y\,dx - x\,dy$. Then, returning to (4), we have

$$\int y\,dx = \frac{1}{2}\int z\,dx + \frac{1}{2}b\,y(b) - \frac{1}{2}a\,y(a)$$

$$= \frac{1}{2}\int [y\,dx - x\,dy] + \frac{1}{2}b\,y(b) - \frac{1}{2}a\,y(a)$$

$$= \frac{1}{2}\int y\,dx - \frac{1}{2}\int x\,dy + \frac{1}{2}b\,y(b) - \frac{1}{2}a\,y(a),$$

which we solve to conclude that $\int y\,dx = b\,y(b) - a\,y(a) - \int x\,dy$.

Again, limits of integration can be inserted to give

$$\int_a^b y\,dx = xy\Big|_a^b - \int_{y(a)}^{y(b)} x\,dy. \tag{6}$$

The geometric validity of (6) is evident in figure 2.7, for $\int_a^b y\,dx$ is the area of the region with vertical strips, whereas $\int_{y(a)}^{y(b)} x\,dy$ is the area of that with horizontal strips. Their sum is clearly the difference in area between the outer rectangle and the small one in the lower left-hand corner. That is,

$$\int_a^b y\,dx + \int_{y(a)}^{y(b)} x\,dy = b\,y(b) - a\,y(a),$$

which can be rearranged into (6).

There is something else about (6) that bears comment: it looks familiar. So it should, because it follows easily from the well-known scheme for integration by parts

$$\int_a^b f(x)g'(x)\,dx = f(x)\,g(x)\Big|_a^b - \int_a^b g(x)\,f'(x)\,dx,$$

if we specify $g(x) = x$ and $f(x) = y$. In that case $g'(x) = 1$ and $f'(x)dx = dy$, and a substitution converts the integration-by-parts formula into the transmutation theorem. After all of Leibniz's convoluted reasoning with its infinitesimals and tangent lines, its similar triangles and wedge-shaped areas—in short, after a most circuitous mathematical journey—we arrive at an instance of integration by parts, a calculus superstar making an early and unexpected entrance onto the stage.

This was intriguing, but Leibniz was not finished. By applying his transmutation theorem to a well-known curve, he discovered the infinite series that still carries his name.

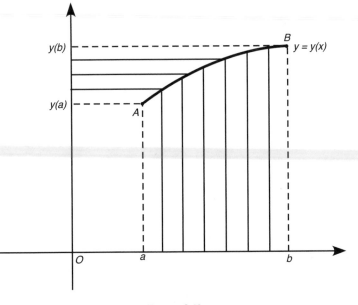

Figure 2.7

THE LEIBNIZ SERIES

Leibniz began with a circular arc. Specifically, he considered a circle of radius 1 and center at $(1, 0)$ and let the curve AB from his general transmutation theorem be the quadrant of this circle shown in figure 2.8. As will become evident momentarily, it was an inspired choice.

The circle's equation is $(x - 1)^2 + y^2 = 1$ or, alternately, $x^2 + y^2 = 2x$. From the geometry of the situation, it is clear that the area beneath the quadrant is $\pi/4$, and so by (1) and (5) we have

$$\frac{\pi}{4} = \int_0^1 y\, dx = \frac{1}{2} xy \Big|_0^1 + \frac{1}{2} \int_0^1 z\, dx, \quad \text{where } z = y - x \frac{dy}{dx}.$$

Using his newly created calculus, Leibniz differentiated the circle's equation to get $2x\, dx + 2y\, dy = 2\, dx$, and so $\dfrac{dy}{dx} = \dfrac{1 - x}{y}$. This led to the simplification

$$z = y - x \frac{dy}{dx} = y - x \left[\frac{1 - x}{y} \right] = \frac{y^2 + x^2 - x}{y} = \frac{2x - x}{y} = \frac{x}{y}.$$

Leibniz's objective was to find an expression for x in terms of the quadratrix z, and so he squared the previous result and again used the equation of the circle to get

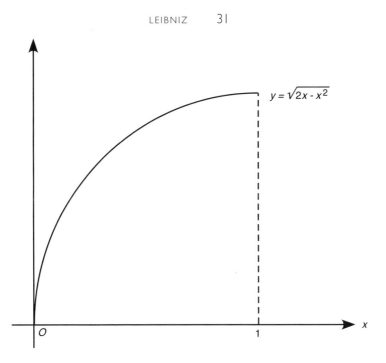

Figure 2.8

$$z^2 = \frac{x^2}{y^2} = \frac{x^2}{2x - x^2} = \frac{x}{2 - x}, \quad \text{which he solved for } x = \frac{2z^2}{1 + z^2}. \quad (7)$$

The challenge was to evaluate $\int_0^1 z \, dx$, the shaded area in figure 2.9. A look at the graph of the quadratrix $z = \sqrt{\dfrac{x}{2 - x}}$ and an observation similar to the one above shows that

$$\int_0^1 z \, dx = \text{Area (shaded region)}$$

$$= \text{Area (square)} - \text{Area (upper region)} = 1 - \int_0^1 x \, dz. \quad (8)$$

Returning to the transmutation theorem, Leibniz combined (7) and (8) as follows:

$$\frac{\pi}{4} = \frac{1}{2} xy \Big|_0^1 + \frac{1}{2} \int_0^1 z \, dx = \frac{1}{2} + \frac{1}{2} \left[1 - \int_0^1 x \, dz \right]$$

$$= 1 - \frac{1}{2} \int_0^1 \frac{2z^2}{1 + z^2} \, dz = 1 - \int_0^1 \frac{z^2}{1 + z^2} \, dz.$$

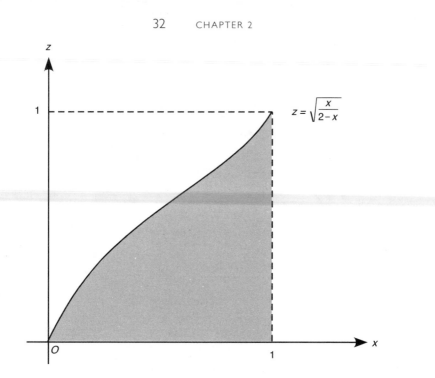

Figure 2.9

He rewrote this last integrand as

$$\frac{z^2}{1+z^2} = z^2\left[\frac{1}{1+z^2}\right] = z^2[1 - z^2 + z^4 - z^6 + \cdots]$$

$$= z^2 - z^4 + z^6 - z^8 + \cdots,$$

where a geometric series has appeared within the brackets. From this, Leibniz concluded that

$$\frac{\pi}{4} = 1 - \int_0^1 [z^2 - z^4 + z^6 - z^8 + \cdots] dz$$

$$= 1 - \left[\frac{z^3}{3} - \frac{z^5}{5} + \frac{z^7}{7} - \frac{z^9}{9} + \cdots\Bigg|_0^1\right] \quad \text{or simply}$$

$$\frac{\pi}{4} = 1 - \frac{1}{3} + \frac{1}{5} - \frac{1}{7} + \frac{1}{9} - \cdots. \tag{9}$$

This is the Leibniz series.

What a wonderful series it is. The terms follow an absolutely trivial pattern: the reciprocals of the odd integers with alternating signs. Yet this innocuous-looking expression sums to, of all things, $\frac{\pi}{4}$. Leibniz recalled that when he first communicated the result to Huygens, he received rave reviews, for "the latter praised it very highly, and when he returned the dissertation said, in the letter that accompanied it, that it would be a discovery always to be remembered among mathematicians" [13].

The significance of this discovery, according to Leibniz, was that "it was now proved for the first time that the area of a circle was exactly equal to a series of rational quantities" [14]. One may quibble with his use of "exactly," but it is hard to argue with his enthusiasm.

He added a curious postscript. By dividing each side of (9) in half and grouping the terms, Leibniz saw that

$$\frac{\pi}{8} = \left(\frac{1}{2} - \frac{1}{6}\right) + \left(\frac{1}{10} - \frac{1}{14}\right) + \left(\frac{1}{18} - \frac{1}{22}\right) + \left(\frac{1}{26} - \frac{1}{30}\right) + \cdots$$

$$= \frac{1}{3} + \frac{1}{35} + \frac{1}{99} + \frac{1}{195} + \cdots$$

$$= \frac{1}{2^2 - 1} + \frac{1}{6^2 - 1} + \frac{1}{10^2 - 1} + \frac{1}{14^2 - 1} + \cdots.$$

In words, this says that if we diminish by 1 the square of every other even number starting with 2 and then add the reciprocals, the sum is $\frac{\pi}{8}$. How strange. One is reminded that formulas from analysis can border on the magical.

The Leibniz series, remarkable as it is, has no value as a numerical approximator of π. The series converges, but it does so with excruciating slowness. One could add the first 300 terms of the Leibniz series and still have π accurate to only a single decimal place. Such dreadful precision would not be worth the effort. However, as we shall see, a related infinite series would, in the hands of Euler, produce a highly efficient scheme for approximating π.

Unquestionably, the Leibniz series is a calculus masterpiece. As is customary when discussing these early results, however, we must offer a few words of caution. For one thing, the transmutation theorem used infinitesimal reasoning. For another, evaluating his series required Leibniz to replace the integral of an infinite sum by the sum of infinitely many integrals, a procedure whose subtleties would be addressed in the centuries to come.

And there was one other problem: Leibniz was not the first to discover this series. The British mathematician James Gregory had found something very similar a few years before. Gregory had, in fact, come upon an expansion for arctangent, namely,

$$\arctan x = x - \frac{x^3}{3} + \frac{x^5}{5} - \frac{x^7}{7} + \cdots,$$

which, for $x = 1$, is the Leibniz series (although Gregory may never have actually made the substitution to convert this to a series of numbers).

Leibniz, a mathematical novice in 1674, was unaware of Gregory's work and believed he had hit upon something new. This in turn led his British counterparts to regard him with some suspicion. To them, Leibniz had a tendency to claim credit for the achievements of others. These suspicions, of course, would be magnified early in the eighteenth century when the British, under the direction of Newton himself, accused Leibniz of outright plagiarism in stealing the calculus. The confusion over the series $\frac{\pi}{4} = 1 - \frac{1}{3} + \frac{1}{5} - \frac{1}{7} + \frac{1}{9} - \cdots$ was seen as an early instance of Leibniz's perfidy.

But even Gregory was not the first down this path. The Indian mathematician Nilakantha, whom we met in the previous chapter, described this series—in verse, no less—in a work called the *Tantrasangraha* [15]. Although it was unknown in Europe during Leibniz's day, this achievement serves as a reminder that mathematics is a universal human enterprise.

The work of Gregory and Nilakantha nothwithstanding, we know that Leibniz's derivation of this series was not theft. He later wrote that in 1674 neither he nor Huygens "nor yet anyone else in Paris had heard anything at all by report concerning the expression of the area of a circle by means of an infinite series of rationals" [16]. The Leibniz series, like the calculus generally, was a personal triumph.

Over the next two decades, the novice would become the master as Leibniz refined, codified, and *published* his ideas on differential and integral calculus. From such beginnings, the subject would grow—indeed, would explode—in the century to come. We continue this story with a look at his two most distinguished followers, the Bernoulli brothers of Switzerland.

CHAPTER 3

The Bernoullis

Jakob Bernoulli   Johann Bernoulli

A scientific revolution often needs more than a founding genius. It may require as well an organizational genius to identify the key ideas, trim off their rough edges, and make them comprehensible to a wider audience. A brilliant architect, after all, may have a vision, but it takes a construction team to turn that vision into a building.

If Newton and Leibniz were the architects of the calculus, it was the Bernoulli brothers, Jakob (1654–1705) and Johann (1667–1748), who did much to build it into the subject we know today. The brothers read Leibniz's original papers from 1684 and 1686 and found them as exhilarating as they were challenging. They grappled with the dense exposition, fleshed out its details, and then, in correspondence with Leibniz and with one another, provided coherence, structure, and terminology. It was Jakob, for instance, who gave us the word "integral" [1]. In their hands, the calculus assumed a form easily recognizable to a student of today, with its basic rules of derivatives, techniques of integration, and solutions of elementary differential equations.

Although excellent mathematicians, the Bernoulli brothers exhibited a personal behavior best described as "unbecoming." Johann, in particular, assumed the combative role of Leibniz's bulldog in the calculus wars with Newton, remaining loyal to his hero, whom he called the "celebrated Leibniz," and going so far as to suggest that not only did Newton fail to invent calculus but he never completely understood it [2]! This was certainly a brazen attack on one of history's greatest mathematicians.

Unfortunately for family harmony, Jakob and Johann were only too happy to do battle with one another. Older brother Jakob, for instance, would refer to Johann as "my pupil," even when the pupil's talents were clearly equal to his own. And, decades after the fact, Johann gleefully recalled solving in a single night a problem that had stumped Jakob for the better part of a year [3].

Their difficult natures notwithstanding, the Bernoullis left deep footprints. Besides his contributions to calculus, Jakob wrote the *Ars conjectandi*, posthumously published in 1713. This work is a classic of probability theory that features a proof of the law of large numbers, a fundamental result that it is sometimes called "Bernoulli's theorem" in his honor [4]. For his part, Johann was the ghostwriter of the world's first calculus text. This came to pass because of an agreement to supply calculus lessons, for a fee, to a French nobleman, the Marquis de l'Hospital (1661–1704). L'Hospital, in turn, assembled and published these in 1696 under the title *Analyse des infiniment petits pour l'intelligence des lignes courbes* (Analysis of the Infinitely Small for the Understanding of Curved Lines). In this work first appeared "l'Hospital's rule," a fixture of differential calculus ever since, although it, like so much of the book, was actually Johann Bernoulli's [5]. In the preface, l'Hospital acknowledged his debt to Bernoulli and Leibniz when he wrote, "I have made free use of their discoveries so that I frankly return to them whatever they please to claim as their own" [6].

The irascible Johann, who indeed claimed the rule, was not satisfied with this gesture and in later years grumbled that l'Hospital had cashed in on the talents of others. Of course it was Bernoulli who (literally) did the cashing in, as math historian Dirk Struik reminded us with this succinct recommendation: "Let the good Marquis keep his elegant rule; he paid for it" [7]. To avoid losing glory a second time, Johann wrote an extensive treatise on integral calculus that was published, under his own name, in 1742 [8].

To get a clearer sense of their mathematical achievements, we shall consider selected works from each brother. We begin with Jakob's divergence proof of the harmonic series, then examine his treatment of some curious convergent series, and conclude with Johann's contributions to what he called the "exponential calculus."

JAKOB AND THE HARMONIC SERIES

Like Newton and Leibniz before him—and so many afterward—Jakob Bernoulli regarded infinite series as a natural pathway into analysis. This was evident in his 1689 work, *Tractatus de seriebus infinitis earumque summa finita* (Treatise on Infinite Series and Their Finite Sums), a state-of-the-art discussion of infinite series as they were understood near the end of the seventeenth century [9]. Jakob considered such familiar series as the geometric, binomial, arctangent, and logarithmic, as well as some previously unexamined ones. In this chapter, we look at two excerpts from the *Tractatus*, the first of which addressed the strange behavior of the harmonic series.

Long before 1689, others had recognized that $1 + \dfrac{1}{2} + \dfrac{1}{3} + \dfrac{1}{4} + \cdots$ diverges to infinity. Nicole Oresme (ca. 1323–1382) devised the proof found in most modern texts, and Pietro Mengoli (1625–1686) came up with an alternate demonstration in 1650 [10]. Leibniz, perhaps unaware of these predecessors, discovered divergence during his early Paris years and informed his British contacts that, in his words, $1 + \dfrac{1}{2} + \dfrac{1}{3} + \dfrac{1}{4} + \cdots = \dfrac{1}{0}$, only to learn from them that he had been scooped once again [11].

So, the divergence of the harmonic series was hardly news. But we may gain insight, not to mention the charm of variety, by following alternate routes to the same end. Jakob Bernoulli's divergence proof, quite different from those of his predecessors, is such an alternative.

He began by comparing two types of progressions that held center stage in his day: the geometric and the arithmetic. The former he described as $A, B, C, D, \ldots$, where $B/A = C/B = D/C$, etc., for example, 2, 1, 1/2, 1/4, ... The latter, he wrote, had the form $A, B, C, D, \ldots$, where $B - A = C - B = D - C$, etc.; an example is 2, 5, 8, 11, ... The modern convention, of course, is to emphasize the common ratio (r) in geometric progressions and the common difference (d) in arithmetic ones, so that we denote a geometric progression by $A, Ar, Ar^2, Ar^3 \ldots$ and an arithmetic one by $A, A + d, A + 2d, A + 3d \ldots$.

As the fourth proposition of his *Tractatus*, Jakob proved a lemma about geometric and arithmetic progressions of positive numbers that begin with the same first *two* terms.

Theorem: If A, B, C, . . . , D, E is a geometric progression of positive numbers with common ratio $r > 1$, and if A, B, F, . . . , G, H is an arithmetic progression of positive numbers also beginning with A and B, then the remaining entries of the geometric progression are greater, term by term, than their arithmetic counterparts.

Proof: Using modern notation, we denote the geometric progression as A, Ar, Ar^2, Ar^3 . . . and the arithmetic one as A, $A + d$, $A + 2d$, $A + 3d$, ... By hypothesis, $Ar = B = A + d$. Because $r > 1$, we have $A(r-1)^2 > 0$, from which it follows that

$$Ar^2 + A > 2Ar,$$

or simply $C + A > 2B = 2(A + d) = A + (A + 2d) = A + F$.

Thus $C > F$; that is, the third term of the geometric series exceeds the third term of the arithmetic one, as claimed. This can be repeated to the fourth, fifth, and indeed to any term down the line.   Q.E.D.

A few propositions later, Jakob proved the following result, stated in characteristic seventeenth century fashion.

Theorem: In any finite geometric progression A, B, C, . . . , D, E, the first term is to the second as the sum of all terms except the last is to the sum of all except the first.

Proof: Once we master the unfamiliar language, this is easily verified because

$$\frac{A}{B} = \frac{A}{Ar} = \frac{A(1 + r + r^2 + \cdots + r^{n-1})}{Ar(1 + r + r^2 + \cdots + r^{n-1})} = \frac{A + Ar + Ar^2 + \cdots + Ar^{n-1}}{Ar + Ar^2 + \cdots + Ar^{n-1} + Ar^n}$$

$$= \frac{A + B + C + \cdots + D}{B + C + \cdots + D + E}.$$ Q.E.D.

Next, Jakob determined the *sum* of a finite geometric progression. Letting $S = A + B + C + \cdots + D + E$ be the sum in question, he applied the previous result to get $\dfrac{A}{B} = \dfrac{S - E}{S - A}$ and then solved for

$$S = \frac{A^2 - BE}{A - B}.$$ (1)

Note that (1) employs the first term (A), the second term (B), and the last term (E) of the finite geometric series, unlike the standard summation formula of today:

$$A + Ar + Ar^2 + \cdots + Ar^k = \frac{A(1 - r^{k+1})}{1 - r},$$

which employs the first term, the number of terms, and the common ratio.

   With these preliminaries aside, we are now ready for Jakob's analysis of the harmonic series. It appeared in the *Tractatus* immediately after a divergence proof credited to Johann [12]. Including his younger brother's work may seem unexpectedly generous, but Jakob rose to the challenge and gave his own alternative. In his words, the goal was to prove that "the sum of the infinite harmonic series $1 + \frac{1}{2} + \frac{1}{3} + \frac{1}{4} + \cdots$ surpasses any given number. Therefore it is infinite" [13].

Theorem: The harmonic series diverges.

Proof: Choosing an arbitrary whole number N, Jakob sought to remove from the beginning of the harmonic series finitely many consecutive terms whose sum is equal to or greater than 1. From what remained, he extracted a finite string of consecutive terms whose sum equals or exceeds another unity. He continued in this fashion until N such strings had been removed, making the sum of the entire harmonic series as least as big as N. Because N was arbitrary, the harmonic series is infinite.

   This procedure, taken almost verbatim from Jakob's original, is fine *provided* we can always remove a finite string of terms whose sum is 1 or more. To complete the argument, Bernoulli had to demonstrate that this is indeed the case. He thus assumed the opposite, stating, "If, after having removed a number of terms, you deny that it is possible for the rest to surpass unity, then let $1/a$ be the first remaining term after the last removal." In other words, for the sake of contradiction,

he supposed that the sum $\frac{1}{a} + \frac{1}{a+1} + \frac{1}{a+2} + \cdots$ remains below 1 no matter how far we carry it. But these denominators $a, a + 1, a + 2, \ldots$ form an arithmetic progression, so Jakob introduced the *geometric* progression beginning with the same first two terms. That is, he considered the geometric progression $a, a+1, C, D, \ldots, K$, where he insisted that we continue until $K \geq a^2$. This is possible because the terms of the progression have a common ratio $r = \dfrac{a+1}{a} > 1$ and thus grow arbitrarily large.

As we saw above, Jakob knew that the terms of the geometric progression exceed those of their arithmetic counterpart, and so, upon taking reciprocals, he concluded that

$$\frac{1}{a} + \frac{1}{a+1} + \frac{1}{a+2} + \cdots > \frac{1}{a} + \frac{1}{a+1} + \frac{1}{C} + \frac{1}{D} + \cdots + \frac{1}{K},$$

where the expression on the left has the same (finite) number of terms as that on the right. He then summed the geometric series using (1) with $A = 1/a$, $B = 1/(a+1)$, and $E = 1/K \leq 1/a^2$ to get

$$\frac{1}{a} + \frac{1}{a+1} + \frac{1}{a+2} + \cdots > \frac{\dfrac{1}{a^2} - \dfrac{1}{a+1}\left[\dfrac{1}{K}\right]}{\dfrac{1}{a} - \dfrac{1}{a+1}} \geq \frac{\dfrac{1}{a^2} - \dfrac{1}{(a+1)a^2}}{\dfrac{1}{a} - \dfrac{1}{a+1}} = 1,$$

a contradiction of his initial assumption. In this way Jakob established that, starting at any point of the harmonic series, a *finite* portion of what remained must sum to one or more.

To complete the proof, he used this scheme to break up the harmonic series as

$$1 + \left(\frac{1}{2} + \frac{1}{3} + \frac{1}{4}\right) + \left(\frac{1}{5} + \frac{1}{6} + \cdots + \frac{1}{25}\right)$$
$$+ \left(\frac{1}{26} + \cdots + \frac{1}{676}\right) + \left(\frac{1}{677} + \cdots + \frac{1}{458329}\right) + \cdots,$$

where each parenthetical expression exceeds 1. The resulting sum can therefore be made greater than any preassigned number, and so the harmonic series diverges.   Q.E.D.

This was a clever argument. Its significance was not lost on Jakob, who emphasized that, "The sum of an infinite series whose final term vanishes is perhaps finite, perhaps infinite" [14]. Of course, no modern mathematician refers to the "final term" of an infinite series, but Jakob's intent is clear: even though the general term of an infinite series shrinks away to zero, this is not sufficient to guarantee convergence. The harmonic series stands as the great example to illustrate this point. So it was for Jakob Bernoulli, and so it remains today.

JAKOB AND HIS FIGURATE SERIES

The harmonic series was of interest because of its bad, that is, divergent, behavior. Of equal interest were well-behaved infinite series having finite sums. Starting with the geometric series and cleverly modifying the outcome, Jakob proceeded until he could calculate the exact values of some nontrivial series. We consider a few of these below.

First he needed the sum of an *infinite* geometric progression. As noted in (1), Bernoulli summed a finite geometric series with the formula

$$A + B + C + \cdots + D + E = \frac{A^2 - BE}{A - B}.$$

As a corollary he observed that, for an infinite geometric progression of positive terms whose common ratio is less than 1, the general term must approach zero. So he simply let his "last" term $E = 0$ to arrive at

$$A + B + C + \cdots + D + \cdots = \frac{A^2}{A - B}. \tag{2}$$

Arithmetic and geometric progressions were not the only patterns familiar to mathematicians of the seventeenth century. So too were the "figurate numbers," families of integers related to such geometrical entities as triangles, pyramids, and cubes. As an example we have the triangular numbers $1, 3, 6, 10, 15, \ldots$, so named because they count the points in the ever-expanding triangles shown in figure 3.1. It is easy to see that the kth triangular number is $1 + 2 + \cdots + k = \dfrac{k(k + 1)}{2} = \dbinom{k + 1}{2}$, where the binomial coefficient is a notation postdating Jakob Bernoulli.

Likewise, the pyramidal numbers are $1, 4, 10, 20, 35, \ldots$, which count the number of cannonballs in pyramidal stacks with triangular bases. It can be shown that the kth pyramidal number is $\dfrac{k(k + 1)(k + 2)}{6} = \dbinom{k + 2}{3}$. Of course, the square numbers $1, 4, 9, 16, 25, \ldots$ and the cubic numbers $1, 8, 27, 64, 125, \ldots$ have geometric significance as well.

Bernoulli's interest in such matters took the following form: he wanted to find the exact sum of an infinite series $\dfrac{a}{A} + \dfrac{b}{B} + \dfrac{c}{C} + \cdots + \dfrac{d}{D} + \cdots,$

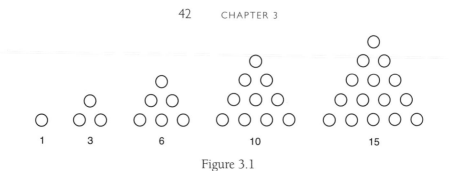

Figure 3.1

where the numerators $a, b, c, \ldots, d, \ldots$ were figurate numbers and the denominators $A, B, C, \ldots, D, \ldots$ constituted a geometric progression. For instance, he wished to evaluate such series as $\displaystyle\sum_{k=1}^{\infty} \frac{\binom{k+2}{3}}{5^k}$ or $\displaystyle\sum_{k=1}^{\infty} \frac{k^3}{2^k}$. These were challenging questions at the time.

Jakob attacked the problem by building from the simple to the complicated—always a good mathematical strategy. Following his arguments, we begin with an infinite series having the natural numbers as numerators and a geometric progression as denominators [15].

Theorem N: If $d > 1$, then $\dfrac{1}{b} + \dfrac{2}{bd} + \dfrac{3}{bd^2} + \dfrac{4}{bd^3} + \dfrac{5}{bd^4} + \cdots = \dfrac{d^2}{b(d-1)^2}$.

Proof: Jakob let $N = \dfrac{1}{b} + \dfrac{2}{bd} + \dfrac{3}{bd^2} + \dfrac{4}{bd^3} + \dfrac{5}{bd^4} + \cdots$ and decomposed it into a sequence of infinite geometric series, each of which he summed by (2):

$$\frac{1}{b} + \frac{1}{bd} + \frac{1}{bd^2} + \frac{1}{bd^3} + \frac{1}{bd^4} + \cdots = \frac{(1/b)^2}{1/b - 1/bd} = \frac{d}{b(d-1)},$$

$$\frac{1}{bd} + \frac{1}{bd^2} + \frac{1}{bd^3} + \frac{1}{bd^4} + \cdots = \frac{(1/bd)^2}{1/bd - 1/bd^2} = \frac{1}{b(d-1)},$$

$$\frac{1}{bd^2} + \frac{1}{bd^3} + \frac{1}{bd^4} + \cdots = \frac{(1/bd^2)^2}{1/bd^2 - 1/bd^3} = \frac{1}{bd(d-1)},$$

$$\frac{1}{bd^3} + \frac{1}{bd^4} + \cdots = \frac{(1/bd^3)^2}{1/bd^3 - 1/bd^4} = \frac{1}{bd^2(d-1)},$$

$$\cdots \qquad = \qquad \cdots \qquad = \qquad \cdots.$$

Upon adding down the columns, he found

$$N = \frac{1}{b} + \frac{2}{bd} + \frac{3}{bd^2} + \frac{4}{bd^3} + \frac{5}{bd^4} + \cdots$$

$$= \frac{d}{b(d-1)} + \frac{1}{b(d-1)} + \frac{1}{bd(d-1)} + \frac{1}{bd^2(d-1)} + \cdots$$

$$= \frac{d}{d-1}\left[\frac{1}{b} + \frac{1}{bd} + \frac{1}{bd^2} + \frac{1}{bd^3} + \cdots\right] = \frac{d}{d-1}\left[\frac{1/b^2}{1/b - 1/bd}\right]$$

$$= \frac{d^2}{b(d-1)^2},$$

because the infinite series in brackets is again geometric.   Q.E.D.

For instance, with $b = 1$ and $d = 7$, we have $1 + \dfrac{2}{7} + \dfrac{3}{49} + \dfrac{4}{343} + \dfrac{5}{2401} + \cdots$
$= \dfrac{7^2}{1 \times 6^2} = \dfrac{49}{36}.$

Next, Jakob put triangular numbers in the numerators.

Theorem T: If $d > 1$, then $T \equiv \dfrac{1}{b} + \dfrac{3}{bd} + \dfrac{6}{bd^2} + \dfrac{10}{bd^3} + \dfrac{15}{bd^4} + \cdots = $
$\dfrac{d^3}{b(d-1)^3}.$

Proof: The trick is to break T into a string of geometric series and exploit
the fact that the kth triangular number is $1 + 2 + 3 + \cdots + k$:

$$\frac{1}{b} + \frac{1}{bd} + \frac{1}{bd^2} + \frac{1}{bd^3} + \frac{1}{bd^4} + \cdots = \frac{(1/b)^2}{1/b - 1/bd} = \frac{d}{b(d-1)},$$

$$\frac{2}{bd} + \frac{2}{bd^2} + \frac{2}{bd^3} + \frac{2}{bd^4} + \cdots = \frac{(2/bd)^2}{2/bd - 2/bd^2} = \frac{2}{b(d-1)},$$

$$\frac{3}{bd^2} + \frac{3}{bd^3} + \frac{3}{bd^4} + \cdots = \frac{(3/bd^2)^2}{3/bd^2 - 3/bd^3} = \frac{3}{bd(d-1)},$$

$$\frac{4}{bd^3} + \frac{4}{bd^4} + \cdots = \frac{(4/bd^3)^2}{4/bd^3 - 4/bd^4} = \frac{4}{bd^2(d-1)},$$

$$\cdots \qquad\qquad = \qquad \cdots \qquad = \qquad \cdots.$$

Adding down the columns gives

$$\frac{1}{b} + \frac{1+2}{bd} + \frac{1+2+3}{bd^2} + \frac{1+2+3+4}{bd^3} + \cdots$$

$$= \frac{d}{b(d-1)} + \frac{2}{b(d-1)} + \frac{3}{bd(d-1)} + \frac{4}{bd^2(d-1)} + \cdots.$$

In other words,

$$T = \frac{d}{d-1}\left[\frac{1}{b} + \frac{2}{bd} + \frac{3}{bd^2} + \frac{4}{bd^3} + \cdots\right]$$

$$= \frac{d}{d-1}N = \frac{d}{d-1} \times \frac{d^2}{b(d-1)^2} = \frac{d^3}{b(d-1)^3},$$

by theorem N.   Q.E.D.

For example, with $b = 2$ and $d = 4$, we have $\dfrac{1}{2} + \dfrac{3}{8} + \dfrac{6}{32} + \dfrac{10}{128} + \dfrac{15}{512} +$

$\cdots = \dfrac{32}{27}.$

Jakob then considered pyramidal numbers in the numerators.

Theorem P: If $d > 1, x$ then $P \equiv \dfrac{1}{b} + \dfrac{4}{bd} + \dfrac{10}{bd^2} + \dfrac{20}{bd^3} + \dfrac{35}{bd^4} + \cdots =$

$$\frac{d^4}{b(d-1)^4}.$$

Proof: This follows easily because

$$P = \left[\frac{1}{b} + \frac{3}{bd} + \frac{6}{bd^2} + \frac{10}{bd^3} + \frac{15}{bd^4} + \cdots\right]$$

$$+ \left[\frac{1}{bd} + \frac{4}{bd^2} + \frac{10}{bd^3} + \frac{20}{bd^4} + \frac{35}{bd^5} + \cdots\right] = T + \frac{1}{d}P.$$

Hence $\left(1 - \dfrac{1}{d}\right)P = T = \dfrac{d^3}{b(d-1)^3}$, and so $P = \dfrac{d^4}{b(d-1)^4}$.   Q.E.D.

As an example, with $b = 5$ and $d = 5$, we have

$$\sum_{k=1}^{\infty} \frac{\binom{k+2}{3}}{5^k} = \frac{1}{5} + \frac{4}{25} + \frac{10}{125} + \frac{20}{625} + \frac{35}{3125} + \cdots = \frac{125}{256}.$$

Jakob finished this part of the *Tractatus* by considering infinite series with the cubic numbers in the numerators and a geometric progression in the denominators.

Theorem C: If $d > 1$, then $C \equiv \dfrac{1}{b} + \dfrac{8}{bd} + \dfrac{27}{bd^2} + \dfrac{64}{bd^3} + \dfrac{125}{bd^4} + \cdots =$

$\dfrac{d^2(d^2 + 4d + 1)}{b(d-1)^4}.$

Proof:

$$C = \left[\frac{1}{b} + \frac{2}{bd} + \frac{3}{bd^2} + \frac{4}{bd^3} + \frac{5}{bd^4} + \cdots \right]$$

$$+ \left[\frac{6}{bd} + \frac{24}{bd^2} + \frac{60}{bd^3} + \frac{120}{bd^4} + \cdots \right]$$

$$= N + \frac{6}{d}\left[\frac{1}{b} + \frac{4}{bd} + \frac{10}{bd^2} + \frac{20}{bd^3} + \frac{35}{bd^4} + \cdots \right] = N + \frac{6}{d}P, \text{ and so}$$

$$C = \frac{d^2}{b(d-1)^2} + \frac{6}{d}\left[\frac{d^4}{b(d-1)^4} \right] = \frac{d^2(d^2 + 4d + 1)}{b(d-1)^4}. \qquad \text{Q.E.D.}$$

When Jakob let $b = 2$ and $d = 2$, he concluded that

$$\sum_{k=1}^{\infty} \frac{k^3}{2^k} = \frac{1}{2} + \frac{8}{4} + \frac{27}{8} + \frac{64}{16} + \frac{125}{32} + \frac{216}{64}$$

$$+ \frac{343}{128} + \frac{512}{256} + \frac{729}{512} + \frac{1000}{1024} + \cdots = 26$$

exactly, surely a strange and nonintuitive result.

After such successes, Jakob Bernoulli may have begun to feel invincible. If he entertained such a notion, he soon had second thoughts, for the series of reciprocals of square numbers, that is, $\displaystyle\sum_{k=1}^{\infty} \frac{1}{k^2}$, resisted all his efforts. He could show, using what we now recognize as the comparison test, that the series converges to some number less than 2, but he was unable to identify it. Swallowing his pride, Jakob included this plea in his *Tractatus*: "If anyone finds and communicates to us that which has thus far eluded our efforts great will be our gratitude" [16].

As we shall see, Bernoulli's challenge went unmet for a generation until finally yielding to one of the greatest analysts of all time.

Jakob Bernoulli was a master of infinite series. His brother Johann, equally gifted, had his own research interests. Among these was what he called the "exponential calculus," which will be our next stop.

JOHANN AND x^x

In a 1697 paper, Johann Bernoulli began with the following general rule: "The differential of a logarithm, no matter how composed, is equal to the differential of the expression divided by the expression" [17]. For instance, $d[\ln(x)] = \dfrac{dx}{x}$ or

$$d[\ln \sqrt{(xx + yy)}] = \frac{1}{2} d[\ln(xx + yy)] = \frac{1}{2}\left[\frac{2xdx + 2ydy}{xx + yy} \right]$$

$$= \frac{xdx + ydy}{xx + yy}.$$

We have retained Bernoulli's original notation for this last expression. At that time in mathematical publishing, higher powers were typeset as they are today, but the quadratic x^2 was often written xx. Also, in the interest of full disclosure, we observe that Bernoulli denoted the natural logarithm of x by lx.

Johann wrote the corresponding integration formula as $\int \dfrac{dx}{x} = lx$. Early in his career he had been seriously confused on this point, believing that $\int \dfrac{dx}{x} = \int x^{-1}dx = \dfrac{1}{0}x^0 = \dfrac{1}{0} \times 1 = \infty$, an overly enthusiastic application of the power rule and one that has yet to be eradicated from the repertoire of beginning calculus students [18]. Fortunately, Johann corrected his error.

With these preliminaries behind him, Johann promised to apply principles "first invented by me" to reap a rich harvest of knowledge "incrementing this new infinitesimal calculus with results not previously found or not widely known" [19]. Perhaps his most interesting example was the curve $y = x^x$, shown in figure 3.2.

For an arbitrary point F on the curve, Johann sought the subtangent, that is, the length of segment LE on the x-axis beneath the tangent line. To do this, he first took logs of both sides: $\ln(y) = \ln(x^x) = x \ln(x)$. He then used his rule to find the differentials:

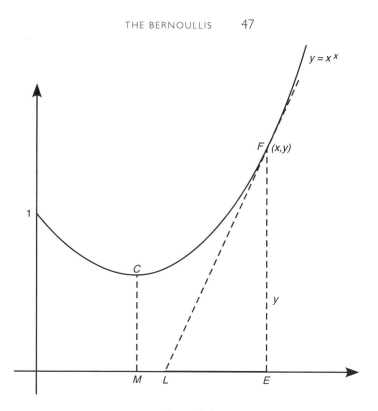

Figure 3.2

$$\frac{1}{y}\,dy = x\frac{dx}{x} + \ln x\,dx = (1+\ln x)dx.$$

But $\dfrac{y}{LE}$ = slope of tangent line $= \dfrac{dy}{dx} = y(1+\ln x)$, and he solved for the

length of the subtangent $LE = \dfrac{y}{y(1+\ln x)} = \dfrac{1}{1+\ln x}$.

Bernoulli next sought the minimum value—what he called the "least of all ordinates"—for the curve. This occurs when the tangent line is horizontal or, equivalently, when the subtangent is infinite. Johann described a somewhat complicated geometric procedure for identifying the value of x for which $1 + \ln x = 0$ [20].

His reasoning was fine, but the form of his answer seems, to modern tastes, less than optimal. Johann was hampered because the introduction of the exponential function still lay decades in the future, so he lacked a notation to express the result simply. We now can solve for $x = 1/e$ and conclude that the minimum value of x^x, that is, the length of segment CM in figure 3.2, is $\left(\dfrac{1}{e}\right)^{1/e} = \dfrac{1}{\sqrt[e]{e}}$, a number roughly equal to 0.6922. This answer, it goes without saying, is by no means obvious.

Johann was just warming up. In another paper from 1697, he tackled a tougher problem: finding the area under his curve $y = x^x$ from $x = 0$ to $x = 1$. That is, he wanted the value of $\int_0^1 x^x\, dx$. Remarkably enough, he found what he was seeking [21].

The argument required two preliminaries. The first he expressed as follows:

If $z = \ln N$, then $N = 1 + z + \dfrac{z^2}{2} + \dfrac{z^3}{2 \times 3} + \dfrac{z^4}{2 \times 3 \times 4} + \cdots$.

Here we recognize the expression for N as the exponential series. If $N = x^x$, then $z = \ln N = x \ln x$, and Johann deduced that

$$x^x = 1 + x\ln x + \frac{x^2(\ln x)^2}{2} + \frac{x^3(\ln x)^3}{2 \times 3} + \frac{x^4(\ln x)^4}{2 \times 3 \times 4} + \cdots. \qquad (3)$$

His objective was to integrate this sum by summing the individual integrals, and for this he needed formulas for $\int x^k(\ln x)^k dx$. He proceeded recursively to generate the table shown on this page.

$$\int dx = x.$$

$$\int x\, l x\, dx = \tfrac{1}{2}\, x x\, l x - \frac{1}{2^2} x x.$$

$$\int x^2 l x^2 dx = \tfrac{1}{3}\, x^3 l x^2 - \frac{2}{3^2} x^3\, l x + \cdot \frac{2}{3^3} x^3$$

$$\int x^3 l x^3 dx = \tfrac{1}{4}\, x^4 l x^3 - \frac{3}{4^2} x^4 l x^2 - \cdot - \frac{3 \cdot 2}{4^3} x^4 l x - \frac{3 \cdot 2}{4^4} x^4.$$

$$\int x^4 l x^4 dx = \tfrac{1}{5}\, x^5 l x^4 - \frac{4}{5^2} x^5 l x^3 + - \frac{4 \cdot 3}{5^3} x^5 l x^2 - \frac{4 \cdot 3 \cdot 2}{5^4} x^5\, l x$$

$$+ \frac{4 \cdot 3 \cdot 2}{5^5} x^5.$$

$$\int x^5 l x^5 dx = \&c.$$

Johann Bernoulli's integral table (1697)

A modern approach would apply integration by parts to prove the reduction formula

$$\int x^m (\ln x)^n \, dx = \frac{1}{m+1} x^{m+1} (\ln x)^n - \frac{n}{m+1} \int x^m (\ln x)^{n-1} \, dx. \qquad (4)$$

For $m = n = 1$, the recursion in (4) gives

$$\int x \ln x \, dx = \frac{1}{2} x^2 \ln x - \frac{1}{2} \int x \, dx = \frac{1}{2} x^2 \ln x - \frac{1}{4} x^2.$$

(Like Bernoulli and other mathematicians of his day, we have ignored "+ C" at the end of the integration formula.) For $m = n = 2$, we have

$$\int x^2 (\ln x)^2 \, dx = \frac{1}{3} x^3 (\ln x)^2 - \frac{2}{3} \int x^2 (\ln x) \, dx$$

$$= \frac{1}{3} x^3 (\ln x)^2 - \frac{2}{3} \left[\frac{1}{3} x^3 \ln x - \frac{1}{3} \int x^2 \, dx \right]$$

$$= \frac{1}{3} x^3 (\ln x)^2 - \frac{2}{9} x^3 \ln x + \frac{2}{27} x^3,$$

where we have also applied (4) with $m = 2$ and $n = 1$.

In this fashion, we replicate Bernoulli's list of integrals. Along with the exponential series in (3), this was the key to solving his curious problem.

Theorem: $\int_0^1 x^x \, dx = 1 - \frac{1}{2^2} + \frac{1}{3^3} - \frac{1}{4^4} + \cdots = \sum_{k=1}^{\infty} \frac{(-1)^{k+1}}{k^k}$.

Proof: By (3), $\int_0^1 x^x \, dx = \int_0^1 \left[1 + x \ln x + \frac{x^2 (\ln x)^2}{2} \right.$

$$\left. + \frac{x^3 (\ln x)^3}{2 \times 3} + \frac{x^4 (\ln x)^4}{2 \times 3 \times 4} + \cdots \right] dx$$

$$= \int_0^1 dx + \int_0^1 x \ln x \, dx + \frac{1}{2} \int_0^1 x^2 (\ln x)^2 \, dx$$

$$+ \frac{1}{2 \times 3} \int_0^1 x^3 (\ln x)^3 \, dx$$

$$+ \frac{1}{2 \times 3 \times 4} \int_0^1 x^4 (\ln x)^4 \, dx + \cdots,$$

where Bernoulli replaced the integral of the series by the series of the integrals without blinking an eye. Using the formulas from his table, he continued:

$$\int_0^1 x^x dx = x\Big|_0^1 + \left(\frac{1}{2}x^2 \ln x - \frac{1}{4}x^2\right)\Big|_0^1$$

$$+ \frac{1}{2}\left(\frac{1}{3}x^3(\ln x)^2 - \frac{2}{9}x^3 \ln x + \frac{2}{27}x^3\right)\Big|_0^1$$

$$+ \frac{1}{2 \times 3}\left(\frac{1}{4}x^4(\ln x)^3 - \frac{3}{16}x^4(\ln x)^2\right.$$

$$\left. + \frac{6}{64}x^4 \ln x - \frac{6}{256}x^4\right)\Big|_0^1$$

$$+ \frac{1}{2 \times 3 \times 4}\left(\frac{1}{5}x^5(\ln x)^4 - \frac{4}{25}x^5(\ln x)^3\right.$$

$$\left. + \frac{12}{125}x^5(\ln x)^2 - \frac{24}{625}x^5 \ln x + \frac{24}{3125}x^5\right)\Big|_0^1 + \cdots.$$

Here he observed that upon substituting $x = 1$, "all terms in which are found lx, or any power . . . of the natural logarithm vanish, insofar as the logarithm of unity is zero" [22]. This is fine, but a modern reader may be puzzled that no mention was made about substituting $x = 0$ to produce indeterminate expressions like $0^m(\ln 0)^n$. Today, we would apply l'Hospital's rule (a most fitting choice!) to show that $\lim_{x \to 0^+} x^m (\ln x)^n = 0$.

In any case, after so many terms had vanished, Bernoulli was left with

$$\int_0^1 x^x dx = 1 - \frac{1}{4} + \frac{1}{2}\left(\frac{2}{27}\right) - \frac{1}{2 \times 3}\left(\frac{6}{256}\right) + \frac{1}{2 \times 3 \times 4}\left(\frac{24}{3125}\right) - \cdots$$

$$= 1 - \frac{1}{4} + \frac{1}{27} - \frac{1}{256} + \frac{1}{3125} - \cdots$$

$$= 1 - \frac{1}{2^2} + \frac{1}{3^3} - \frac{1}{4^4} + \frac{1}{5^5} - \cdots. \qquad \text{Q.E.D.}$$

It is quite remarkable that this series gives the area beneath the curve $y = x^x$ over the unit interval. Beyond its splendid symmetry and immediate visual appeal, it has another attribute not lost on Johann. He noted, "This wonderful series converges so rapidly that the tenth term contributes only a thousandth of a millionth part of unity to the sum" [23]. To be sure, it

takes only a handful of terms to calculate $\int_0^1 x^x dx \approx 0.7834305107$ accurately to ten places.

As the examples in this chapter should make clear, Jakob and Johann Bernoulli were worthy disciples of Gottfried Wilhelm Leibniz. In their hands, his calculus became, as we might say today, "user-friendly." The brothers left the subject in a more sophisticated yet much more understandable state than they found it.

And Johann had one other legacy. In the 1720s, he mentored a young Swiss student of almost limitless promise. The student's name was Leonhard Euler, and we sample his work next.

CHAPTER 4

Euler

Leonhard Euler

In any accounting of history's greatest mathematicians, Leonhard Euler (1707–1783) stands tall. With broad and inexhaustible interests, he revolutionized mathematics, extending the boundaries of such well-established subdisciplines as number theory, algebra, and geometry even while giving birth to new ones like graph theory, the calculus of variations, and the theory of partitions. When in 1911 scholars began publishing his collected works, the *Opera omnia*, they faced a daunting challenge. Today, after more than seventy volumes and 25,000 pages in print, the task is not yet complete. This enormous publishing project, consuming the better part of a century, bears witness to a mathematical force of nature.

That force was especially evident in analysis. Among Euler's collected works are *eighteen* thick volumes and nearly 9000 pages on the subject. These include landmark textbooks on functions (1748), differential calculus (1755), and integral calculus (1768), as well as dozens of papers on topics ranging from differential equations to infinite series to elliptic integrals. As a consequence, Euler has been described as "analysis incarnate" [1].

It is impossible to do justice to these contributions in a short chapter. Rather, we have selected five topics to illustrate the sweep of Euler's achievements. We begin with an example from elementary calculus, featuring the bold—some may say reckless—approach so characteristic of his work.

A DIFFERENTIAL FROM EULER

In his text *Institutiones calculi differentialis* of 1755, Euler presented the familiar formulas of differential calculus [2]. These depended upon the notion of "infinitely small quantities," which he characterized as follows:

> There is no doubt that any quantity can be diminished until it all but vanishes and then goes to nothing. But an infinitely small quantity is nothing but a vanishing quantity, and so it is really equal to 0. . . . There is really not such a great mystery lurking in this idea as some commonly think and thus have rendered the calculus of the infinitely small suspect to so many. [3]

For Euler, the differential dx was zero: nothing more, nothing less—in short, nothing at all. The expressions x and $x + dx$ were therefore equal and could be interchanged as the situation required. He observed that "the infinitely small vanishes in comparison with the finite and hence can be neglected" [4]. Moreover, powers like $(dx)^2$ or $(dx)^3$ are infinitely *smaller* than the infinitely small dx and likewise can be jettisoned at will.

It was often the *ratio* of differentials that Euler sought, and determining this ratio, which amounted to assigning a value to 0/0, was the mission of calculus. As he put it, "the whole force of differential calculus is concerned with the investigation of the ratios of any two infinitely small quantities" [5].

As an illustration, we consider his treatment of the function $y = \sin x$. Euler began with Newton's series (where we employ the modern "factorial" notation):

$$\sin z = z - \frac{z^3}{3!} + \frac{z^5}{5!} - \frac{z^7}{7!} + \cdots \text{ and}$$

$$\cos z = 1 - \frac{z^2}{2!} + \frac{z^4}{4!} - \frac{z^6}{6!} + \cdots. \tag{1}$$

Substituting the differential dx for z, he reasoned that

$$\sin(dx) = dx - \frac{(dx)^3}{3!} + \frac{(dx)^5}{5!} - \frac{(dx)^7}{7!} + \cdots \quad \text{and}$$

$$\cos(dx) = 1 - \frac{(dx)^2}{2!} + \frac{(dx)^4}{4!} - \frac{(dx)^6}{6!} + \cdots .$$

Because the higher powers of the differential are insignificant compared to dx or to constants, these series reduced to

$$\sin(dx) = dx \quad \text{and} \quad \cos(dx) = 1. \tag{2}$$

In the equation $y = \sin x$, Euler replaced x by $x + dx$ and y by $y + dy$ (which for him changed nothing) and employed the identity $\sin(\alpha + \beta) = \sin \alpha \cos \beta + \cos \alpha \sin \beta$ and (2) to get

$$y + dy = \sin(x + dx) = \sin x \cos(dx) + \cos x \sin(dx) = \sin x + (\cos x)dx.$$

Subtracting $y = \sin x$ from both sides, he was left with $dy = \sin x + (\cos x)dx - y = (\cos x)\, dx$, which he turned into a verbal recipe: "the differential of the sine of any arc is equal to the product of the differential of the arc and the cosine of the arc" [6]. It follows that the *ratio* of these differentials—what we, of course, call the derivative—is $\dfrac{dy}{dx} = \dfrac{(\cos x)dx}{dx} = \cos x$. Nothing to it!

AN INTEGRAL FROM EULER

Euler was one of history's foremost integrators, and the more bizarre the integrand, the better. His works, particularly volumes 17, 18, and 19 of the *Opera omnia*, are filled with such nontrivial examples as [7]:

$$\int_0^1 \frac{(\ln x)^5}{1+x}\, dx = -\frac{31\pi^6}{252},$$

$$\int_0^\infty \frac{\sin x}{x}\, dx = \frac{\pi}{2},$$

$$\int_0^1 \frac{\sin(p \ln x) \cdot \cos(q \ln x)}{\ln x}\, dx = \frac{1}{2} \arctan\left(\frac{2p}{1 - p^2 + q^2}\right).$$

This last features a particularly rich mixture of transcendental functions.

As our lone representative, we consider Euler's evaluation of $\int_0^1 \frac{\sin(\ln x)}{\ln x} dx$ [8]. To begin, he employed a favorite strategy: introduce an infinite series when possible. From (1), he knew that

$$\frac{\sin(\ln x)}{\ln x} = \frac{\ln x - \frac{(\ln x)^3}{3!} + \frac{(\ln x)^5}{5!} - \frac{(\ln x)^7}{7!} + \cdots}{\ln x}$$

$$= 1 - \frac{(\ln x)^2}{3!} + \frac{(\ln x)^4}{5!} - \frac{(\ln x)^6}{7!} + \cdots.$$

Replacing the integral of the infinite series by the infinite series of integrals, he continued:

$$\int_0^1 \frac{\sin(\ln x)}{\ln x} dx = \int_0^1 dx - \frac{1}{3!}\int_0^1 (\ln x)^2 dx + \frac{1}{5!}\int_0^1 (\ln x)^4 dx$$

$$- \frac{1}{7!}\int_0^1 (\ln x)^6 dx + \cdots. \qquad (3)$$

Integrals of the form $\int_0^1 (\ln x)^n dx$ are reminiscent of Johann Bernoulli's formulas from the previous chapter, and Euler instantly spotted their recursive pattern:

$$\int_0^1 (\ln x)^2 dx = \left[x(\ln x)^2 - 2x\ln x + 2x \right]\Big|_0^1 = 2 = 2!,$$

$$\int_0^1 (\ln x)^4 dx = \left[x(\ln x)^4 - 2x(\ln x)^3 + 12x(\ln x)^2 \right.$$

$$\left. -24x\ln x + 24x \right]\Big|_0^1 = 24 = 4!,$$

$$\int_0^1 (\ln x)^6 dx = 720 = 6!, \text{ and so on}.$$

As noted in the previous chapter, $\lim_{x \to 0^+} x(\ln x)^n = 0$, which explains the disappearance of terms arising from substituting zero for x in these anti-derivatives.

When Euler applied this pattern to (3), he found that

$$\int_0^1 \frac{\sin(\ln x)}{\ln x} dx = 1 - \frac{1}{3!}[2] + \frac{1}{5!}[24] - \frac{1}{7!}[720] + \cdots$$

$$= 1 - \frac{1}{3} + \frac{1}{5} - \frac{1}{7} + \frac{1}{9} - \cdots.$$

This, of course, is the Leibniz series from chapter 2, so Euler finished in style:

$$\int_0^1 \frac{\sin(\ln x)}{\ln x}\, dx = \frac{\pi}{4}.$$

The derivation shows that Euler—like Newton, Leibniz, and the Bernoullis before him—was a spectacular (and fearless!) manipulator of infinite series. In fact, one could argue, based on the mathematicians seen thus far, that a high comfort level in working with infinite series *defined* an analyst in these early days.

The appearance of π in the integral above leads us directly to the next topic: Euler's techniques for approximating this famous number.

EULER'S ESTIMATION OF π

By definition, π is the ratio of a circle's circumference to its diameter. From ancient times, people recognized that the ratio was constant from one circle to another, but attaching a numerical value to this constant has kept mathematicians busy for centuries.

As is well known, Archimedes approximated π by inscribing (and circumscribing) regular polygons in (and about) a circle and then using the polygons' perimeters to estimate the circle's circumference. He began with regular inscribed and circumscribed hexagons and, upon doubling the number of sides to 12, to 24, to 48, and finally to 96, he showed that "the ratio of the circumference of any circle to its diameter is less than $3\frac{1}{7}$ but greater than $3\frac{10}{71}$" [9]. To two-place accuracy, this means $\pi \approx 3.14$.

Subsequent mathematicians, whose number system was computationally simpler than that available in classical Greece, exploited his idea. In 1579, François Viète (1540–1603) found π accurately to nine places using polygons with $6 \times 2^{16} = 393{,}216$ sides. This geometrical approach reached a kind of zenith (or nadir) in the work of Ludolph van Ceulen (1540–1610), who used regular 2^{62}-gons to calculate π to 35 decimal places in a phenomenal display of applied tedium that reportedly consumed the better part of his life [10].

Unfortunately, each new approximation in this process required taking a new square root. The estimate of π generated by Archimedes' inscribed 96-gon was

$$48\sqrt{2 - \sqrt{2 + \sqrt{2 + \sqrt{2 + \sqrt{3}}}}}\,,$$

an expression that is a treat to the eye but a nightmare to the pencil. Yet after these five square root extractions, we have only two-place accuracy. Worse was Viète's nesting of seventeen square roots for his nine places of accuracy, and unthinkably awful was Ludolph's approximation featuring five dozen nested radicals, each calculated to thirty-five places— by hand! Euler compared such work unfavorably to the labors of Hercules [11].

Fortunately, there was another way. As we mentioned in chapter 2, James Gregory discovered the infinite series for arctangent:

$$\arctan x = x - \frac{x^3}{3} + \frac{x^5}{5} - \frac{x^7}{7} + \cdots. \tag{4}$$

For $x = 1$, this becomes Leibniz's series $\frac{\pi}{4} = \arctan(1) = 1 - \frac{1}{3} + \frac{1}{5} - \frac{1}{7} + \frac{1}{9} - \cdots$, which, as we observed, is of no value in approximating π because of its glacial rate of convergence.

However, if we substitute a value of x closer to zero, the convergence is more rapid. For instance, letting $x = \frac{1}{\sqrt{3}}$ in (4), we get

$$\frac{\pi}{6} = \arctan\left(\frac{1}{\sqrt{3}}\right)$$

$$= \frac{1}{\sqrt{3}} - \frac{1}{(3\sqrt{3}) \times 3} + \frac{1}{(9\sqrt{3}) \times 5} - \frac{1}{(27\sqrt{3}) \times 7} + \cdots,$$

so that

$$\pi = \frac{6}{\sqrt{3}}\left[1 - \frac{1}{3 \times 3} + \frac{1}{9 \times 5} - \frac{1}{27 \times 7} + \cdots\right].$$

This is an improvement over the Leibniz series because its denominators are growing much faster. On the other hand, $\frac{1}{\sqrt{3}} \approx 0.577$, which is not all that small, and this series involves a square root that itself would have to be approximated.

For a mathematician of the eighteenth century, the ideal formula would use Gregory's infinite series with a value of x quite close to zero while avoiding square roots altogether. This is precisely what Euler described in a

1779 paper [12]. His key observation, which at first glance looks like a typographical error, was that

$$\pi = 20 \arctan(1/7) + 8 \arctan(3/79). \tag{5}$$

Improbable though it may seem, this is an *equation*, not an estimate. Here is how Euler proved it.

He started with the identity $\tan(\alpha - \beta) = \dfrac{\tan \alpha - \tan \beta}{1 + (\tan \alpha)(\tan \beta)}$, which can be recast as $\alpha - \beta = \arctan\left[\dfrac{\tan \alpha - \tan \beta}{1 + (\tan \alpha)(\tan \beta)}\right]$. Euler let $\tan \alpha = \dfrac{x}{y}$ and $\tan \beta = \dfrac{z}{w}$ to get

$$\arctan\left(\frac{x}{y}\right) - \arctan\left(\frac{z}{w}\right) = \arctan\left[\frac{\dfrac{x}{y} - \dfrac{z}{w}}{1 + \left(\dfrac{x}{y}\right)\left(\dfrac{z}{w}\right)}\right],$$

or simply

$$\arctan\left(\frac{x}{y}\right) = \arctan\left(\frac{z}{w}\right) + \arctan\left[\frac{xw - yz}{yw + xz}\right]. \tag{6}$$

He then substituted a string of cleverly chosen rationals. First, Euler set $x = y = z = 1$ and $w = 2$ in (6) to get $\dfrac{\pi}{4} = \arctan(1) = \arctan\left(\dfrac{1}{2}\right) + \arctan\left(\dfrac{1}{3}\right)$, so that

$$\pi = 4 \arctan\left(\frac{1}{2}\right) + 4 \arctan\left(\frac{1}{3}\right). \tag{7}$$

He could have stopped there, using (7) to approximate π via Gregory's arctangent series, but the input values of 1/2 and 1/3 were too large to give the rapid convergence he desired. Instead, Euler returned to (6) with $x = 1$, $y = 2, z = 1$, and (for reasons not immediately apparent) $w = 7$. This led to

$$\arctan(1/2) = \arctan(1/7) + \arctan(5/15) = \arctan(1/7) + \arctan(1/3),$$

which, when substituted into (7), gave the new expression

$$\pi = 4[\arctan(1/7) + \arctan(1/3)] + 4 \arctan(1/3)$$
$$= 4 \arctan(1/7) + 8 \arctan(1/3). \tag{8}$$

Next, Euler chose $x = 1$, $y = 3$, $z = 1$, and $w = 7$ to conclude from (6) that $\arctan(1/3) = \arctan(1/7) + \arctan(2/11)$. This he substituted into (8) to get

$$\pi = 12 \arctan(1/7) + 8 \arctan(2/11). \tag{9}$$

In a final iteration of (6), Euler let $x = 2$, $y = 11$, $z = 1$, and $w = 7$ so that $\arctan(2/11) = \arctan(1/7) + \arctan(3/79)$, which in turn transformed (9) into the peculiar result stated in (5):

$$\pi = 12 \arctan(1/7) + 8 [\arctan(1/7) + \arctan(3/79)]$$
$$= 20 \arctan(1/7) + 8 \arctan(3/79).$$

This expression for π is admirably suited to the arctangent series in (4), for it is free of square roots and uses the relatively small numbers 1/7 and 3/79 to produce rapid convergence. With only six terms from each series, we calculate

$$\pi = 20 \arctan\left(\frac{1}{7}\right) + 8 \arctan\left(\frac{3}{79}\right)$$
$$\approx 20\left[\frac{1}{7} - \frac{(1/7)^3}{3} + \frac{(1/7)^5}{5} - \frac{(1/7)^7}{7} + \frac{(1/7)^9}{9} - \frac{(1/7)^{11}}{11}\right]$$
$$+ 8\left[\frac{3}{79} - \frac{(3/79)^3}{3} + \frac{(3/79)^5}{5} - \frac{(3/79)^7}{7}\right.$$
$$\left. + \frac{(3/79)^9}{9} - \frac{(3/79)^{11}}{11}\right]$$
$$\approx 3.14159265357.$$

Here, a dozen fractions provide an estimate of π accurate to two parts in a hundred billion, a better approximation than Viète obtained by extracting seventeen nested square roots. In fact, Euler claimed to have used such techniques to approximate π to twenty places, "and all this calculation consumed about an hour of work" [13].

Recalling the lifetime that poor Ludolph devoted to his bewildering tangle of square roots, one is tempted to change Euler's nickname to "efficiency incarnate."

SPECTACULAR SUMS

In this section we shall see how Euler, by analyzing a single situation, was able to find the *exact* values of

$$\sum_{k=1}^{\infty} \frac{(-1)^{k+1}}{2k-1} = 1 - \frac{1}{3} + \frac{1}{5} - \frac{1}{7} + \cdots \text{ (Leibniz's series)},$$

$$\sum_{k=1}^{\infty} \frac{1}{k^2} = 1 + \frac{1}{4} + \frac{1}{9} + \frac{1}{16} + \cdots \text{ (Jakob Bernoulli's challenge)},$$

$$\sum_{k=1}^{\infty} \frac{(-1)^{k+1}}{(2k-1)^3} = 1 - \frac{1}{27} + \frac{1}{125} - \frac{1}{343} + \cdots, \text{ and many more.}$$

By unifying these sums under one theory, Euler cemented his reputation as one of history's great series manipulators.

The story begins with a result from his 1748 text, *Introductio in analysin infinitorum*.

Lemma: If $P(x) = 1 + Ax + Bx^2 + Cx^3 + \cdots = (1 + \alpha_1 x)(1 + \alpha_2 x)$ $(1 + \alpha_3 x) \ldots$, then

$$\sum \alpha_k = A,$$
$$\sum \alpha_k^2 = A^2 - 2B,$$
$$\sum \alpha_k^3 = A^3 - 3AB + 3C,$$
$$\sum \alpha_k^4 = A^4 - 4A^2B + 4AC + 2B^2 - 4D, \text{ and so on,}$$

whether these factors be "finite or infinite in number" [14].

Proof: Euler observed that such formulas were "intuitively clear," but promised a rigorous argument using differential calculus. This appeared in a 1750 paper on the theory of equations [15].

Before proving the lemma, we should clarify its meaning. Setting $0 = P(x) = (1 + \alpha_1 x)(1 + \alpha_2 x)(1 + \alpha_3 x) \ldots$, we solve for $x = -1/\alpha_1$, $-1/\alpha_2$, $-1/\alpha_3, \ldots$. The lemma thus connects the coefficients A, B, C, $\ldots$ in the expression for P to the negative reciprocals of the solutions to $P(x) = 0$. In this light, the result seems to be primarily an *algebraic* one.

But Euler, the great analyst, saw it differently. He started by taking logarithms:

$$\ln[P(x)] = \ln[1 + Ax + Bx^2 + Cx^3 + \cdots]$$
$$= \ln[(1 + \alpha_1 x)(1 + \alpha_2 x)(1 + \alpha_3 x) \ldots]$$
$$= \ln(1 + \alpha_1 x) + \ln(1 + \alpha_2 x) + \ln(1 + \alpha_3 x) + \cdots.$$

Then, making good on his promise to use calculus, he differentiated both sides to get

$$\frac{A + 2Bx + 3Cx^2 + 4Dx^3 + \cdots}{1 + Ax + Bx^2 + Cx^3 + \cdots} = \frac{\alpha_1}{1 + \alpha_1 x} + \frac{\alpha_2}{1 + \alpha_2 x} + \frac{\alpha_3}{1 + \alpha_3 x} + \cdots. \quad (10)$$

It was evident to Euler that each fraction $\dfrac{\alpha_k}{1 + \alpha_k x}$ on the right-hand side was the sum of an infinite geometric series with first term α_k and common ratio $-\alpha_k x$. That is,

$$\frac{\alpha_1}{1 + \alpha_1 x} = \alpha_1 - \alpha_1^2 x + \alpha_1^3 x^2 - \alpha_1^4 x^3 + \cdots,$$

$$\frac{\alpha_2}{1 + \alpha_2 x} = \alpha_2 - \alpha_2^2 x + \alpha_2^3 x^2 - \alpha_2^4 x^3 + \cdots,$$

$$\frac{\alpha_3}{1 + \alpha_3 x} = \alpha_3 - \alpha_3^2 x + \alpha_3^3 x^2 - \alpha_3^4 x^3 + \cdots, \text{ and so on}.$$

Upon adding down the columns of this array and summing like powers of α_k, he rewrote equation (10) as

$$\frac{A + 2Bx + 3Cx^2 + 4Dx^3 + \cdots}{1 + Ax + Bx^2 + Cx^3 + \cdots}$$
$$= \sum \alpha_k - \left(\sum \alpha_k^2\right)x + \left(\sum \alpha_k^3\right)x^2 - \left(\sum \alpha_k^4\right)x^3 + \cdots.$$

This he cross-multiplied and expanded to get

$$A + 2Bx + 3Cx^2 + 4Dx^3 + \cdots$$
$$= [1 + Ax + Bx^2 + Cx^3 + \cdots]$$
$$\times \left[\sum \alpha_k - \left(\sum \alpha_k^2 \right) x + \left(\sum \alpha_k^3 \right) x^2 - \left(\sum \alpha_k^4 \right) x^3 + \cdots \right]$$
$$= \sum \alpha_k + \left[A \sum \alpha_k - \sum \alpha_k^2 \right] x + \left[B \sum \alpha_k - A \sum \alpha_k^2 + \sum \alpha_k^3 \right] x^2$$
$$+ \left[C \sum \alpha_k - B \sum \alpha_k^2 + A \sum \alpha_k^3 - \sum \alpha_k^4 \right] x^3 + \cdots.$$

From here, Euler equated coefficients of like powers of x and so determined $\sum \alpha_k^m$ recursively:

(a) $\sum \alpha_k = A,$

(b) $\left[A \sum \alpha_k - \sum \alpha_k^2 \right] = 2B,$ and so
$$\sum \alpha_k^2 = \left[A \sum \alpha_k - 2B \right] = A^2 - 2B,$$

(c) $B \sum \alpha_k - A \sum \alpha_k^2 + \sum \alpha_k^3 = 3C,$ and so
$$\sum \alpha_k^3 = A \sum \alpha_k^2 - B \sum \alpha_k + 3C$$
$$= A[A^2 - 2B] - AB + 3C = A^3 - 3AB + 3C,$$

(d) $C \sum \alpha_k - B \sum \alpha_k^2 + A \sum \alpha_k^3 - \sum \alpha_k^4 = 4D,$ and so
$$\sum \alpha_k^4 = A^4 - 4A^2B + 4AC + 2B^2 - 4D.$$

The process can be continued at will. In this way, by combining logarithms, derivatives, and geometric series, Euler proved his "intuitively clear" formulas!   Q.E.D.

To demonstrate their relevance, he considered the general expression $P(x) = \cos\left(\dfrac{\pi}{2n} x \right) + \left(\tan \dfrac{m\pi}{2n} \right) \sin\left(\dfrac{\pi}{2n} x \right)$ although we here restrict our attention to the case where $m = 1$ and $n = 2$ [16]. That is, we consider

$$P(x) = \cos\left(\frac{\pi}{4} x \right) + \left(\tan \frac{\pi}{4} \right) \sin\left(\frac{\pi}{4} x \right) = \cos\left(\frac{\pi}{4} x \right) + \sin\left(\frac{\pi}{4} x \right).$$

To apply the lemma, we must write P as an infinite series and as an infinite product of factors of the form $(1 + \alpha_k x)$, where $-1/\alpha_k$ is a root of $P(x) = 0$. The former is easy, for we need only shuffle together the series for cosine and sine from (1) to get

$$P(x) = 1 + \frac{\pi}{4} x - \frac{\pi^2}{4^2 \cdot 2!} x^2 - \frac{\pi^3}{4^3 \cdot 3!} x^3$$
$$+ \frac{\pi^4}{4^4 \cdot 4!} x^4 + \frac{\pi^5}{4^5 \cdot 5!} x^5 - \cdots.$$

We thus identify the coefficients from the lemma as

$$A = \pi/4,\ B = -\pi^2/32,\ C = -\pi^3/384,\ D = \pi^4/6144, \ldots.$$

On the other hand, setting $0 = P(x) = \cos\left(\dfrac{\pi}{4} x\right) + \sin\left(\dfrac{\pi}{4} x\right)$ leads to $\tan\dfrac{\pi}{4} x = -1$, whose roots are $x = -1,\ 3,\ -5,\ 7,\ -9, \ldots$. The negative reciprocals of these roots will be the α_k from the lemma, so that

$$\alpha_1 = 1,\ \alpha_2 = -1/3,\ \alpha_3 = 1/5,\ \alpha_4 = -1/7,\ \alpha_5 = 1/9, \ldots.$$

At last Euler could reap his rewards. According to the lemma, $\sum \alpha_k = A$ and so $1 - \dfrac{1}{3} + \dfrac{1}{5} - \dfrac{1}{7} + \dfrac{1}{9} - \cdots = \dfrac{\pi}{4}$. Here we have the Leibniz series making a return appearance. Note that in contrast to Leibniz's complicated, geometric derivation from chapter 2, Euler's was purely analytic with no evident triangles, curves, or graphs.

The lemma's second relationship was $\sum \alpha_k^2 = A^2 - 2B$, which for our specific function P provides the sum of reciprocals of the odd squares:

$$1 + \frac{1}{9} + \frac{1}{25} + \frac{1}{49} + \frac{1}{81} - \cdots = \left(\frac{\pi}{4}\right)^2 - 2\left(-\frac{\pi^2}{32}\right) = \frac{\pi^2}{8}.$$

From this, Euler could easily answer Bernoulli's question about the sum of the reciprocals of *all* the squares, because

$$1 + \frac{1}{4} + \frac{1}{9} + \frac{1}{16} + \frac{1}{25} + \frac{1}{36} + \frac{1}{49} + \cdots$$
$$= \left(1 + \frac{1}{9} + \frac{1}{25} + \frac{1}{49} + \frac{1}{81} + \cdots\right) + \frac{1}{4}\left(1 + \frac{1}{4} + \frac{1}{9} + \frac{1}{16} + \frac{1}{25} + \cdots\right).$$

It follows that $\dfrac{3}{4}\left(1+\dfrac{1}{4}+\dfrac{1}{9}+\dfrac{1}{16}+\dfrac{1}{25}+\dfrac{1}{36}+\dfrac{1}{49}+\cdots\right)=\left(1+\dfrac{1}{9}+\dfrac{1}{25}+\right.$

$\left.\dfrac{1}{49}+\dfrac{1}{81}+\cdots\right)=\dfrac{\pi^2}{8}$, and so $1+\dfrac{1}{4}+\dfrac{1}{9}+\dfrac{1}{16}+\dfrac{1}{25}+\dfrac{1}{36}+\dfrac{1}{49}+\cdots=$

$\dfrac{4}{3}\times\dfrac{\pi^2}{8}=\dfrac{\pi^2}{6}$. The resolution of Bernoulli's challenge was another feather in Euler's feather-laden hat.

The next equation from the lemma, $\sum \alpha_k^3 = A^3 - 3AB + 3C$, yielded the alternating series:

$$1-\frac{1}{27}+\frac{1}{125}-\frac{1}{343}+\frac{1}{729}-\cdots$$

$$=\left(\frac{\pi}{4}\right)^3-3\left(\frac{\pi}{4}\right)\left(-\frac{\pi^2}{32}\right)+3\left(-\frac{\pi^3}{384}\right)=\frac{\pi^3}{32}.$$

And on he went, using the lemma repeatedly to derive such formulas as $\displaystyle\sum_{k=1}^{\infty}\frac{1}{k^4}=\frac{\pi^4}{90}$ and $\displaystyle\sum_{k=1}^{\infty}\frac{(-1)^{k+1}}{(2k-1)^5}=\frac{5\pi^5}{1536}$ and many more. This spectacular achievement calls to mind Ivor Grattan-Guinness's observation that "Euler was the high priest of sum-worship, for he was cleverer than anyone else at inventing unorthodox methods of summation" [17]. It goes without saying that the high priest was agnostic about subtle convergence questions accompanying his proof. Such matters would have to await the next century.

One other striking fact leaps off the page. Although Euler had evaluated expressions like $\displaystyle\sum_{k=1}^{\infty}\frac{1}{k^2}$, and $\displaystyle\sum_{k=1}^{\infty}\frac{1}{k^4}$, he did not explicitly sum $\displaystyle\sum_{k=1}^{\infty}\frac{1}{k^3}$ or other series with odd exponents. The value of such quantities, wrote Euler, "can be expressed neither by logarithms nor by the circular periphery π, nor can a value be assigned by any other finite means" [18]. At one point, stumped by this vexing problem, an apparently frustrated Euler conceded that it would be "to no purpose" for him to investigate further [19]. It says something for his analytic intuition that to this day the nature of these odd-powered series remains far from clear. One suspects that if Euler failed to find a simple solution, it does not exist.

We conclude with one other significant contribution to analysis: Euler's ideas on extending factorials to noninteger inputs.

THE GAMMA FUNCTION

An interesting mathematical exercise is to *interpolate* a formula involving whole numbers. That is, we seek an expression, defined across a larger domain, that agrees with the original formula when the input is a positive integer.

By way of clarification, consider the following example discussed by Philip Davis in an article on the origins of the gamma function [20]. For any positive integer n, we let $S(n) = 1 + 2 + 3 + \cdots + n$ be the sum of the first n whole numbers. Clearly, $S(4) = 1 + 2 + 3 + 4 = 10$. It would make no sense, however, to talk about the sum of the first *four-and-a-quarter* numbers.

To make that leap, we introduce a function T defined for all real x by $T(x) = \dfrac{x(x+1)}{2}$. Here T interpolates S, for when n is a whole number, $S(n) = 1 + 2 + 3 + \cdots + n = \dfrac{n(n+1)}{2} = T(n)$. But now we *can* evaluate $T(4.25) = 11.15625$. In this way, the function T "fills the gaps" in our representation of S, or, as Davis put it, "the formula extends the scope of the original problem to values of the variable other than those for which it was originally defined."

In fact, this is what Newton did with his generalized binomial expansion. Rather than restrict himself to whole number powers of $(1 + x)^n$, he dealt with fractional or negative exponents in a way that matched, that is, interpolated, the familiar situation when n was a positive integer.

In 1729, the ever-curious Euler took up an analogous challenge for the *product* of the first n whole numbers. That is, he sought a formula defined for all positive real numbers that agreed with $1 \cdot 2 \cdot 3 \cdot \ldots \cdot n$ when the input n was a positive integer. To use modern terminology, Euler sought to interpolate the factorial.

His first solution appeared in a letter to Christian Goldbach from October of 1729 [21]. There, he proposed the bizarre-looking infinite product

$$\frac{1 \cdot 2^x}{1+x} \times \frac{2^{1-x} \cdot 3^x}{2+x} \times \frac{3^{1-x} \cdot 4^x}{3+x} \times \frac{4^{1-x} \cdot 5^x}{4+x} \times \cdots. \tag{11}$$

At different times, Euler denoted this expression by $\Delta(x)$ and by $[x]$. For the remainder of the chapter, we shall use the latter. From (11) one sees that

$$[1] = \frac{1 \cdot 2}{2} \times \frac{1 \cdot 3}{3} \times \frac{1 \cdot 4}{4} \times \frac{1 \cdot 5}{5} \times \cdots = 1,$$

$$[2] = \frac{1 \cdot 2 \cdot 2}{3} \times \frac{3 \cdot 3}{2 \cdot 4} \times \frac{4 \cdot 4}{3 \cdot 5} \times \frac{5 \cdot 5}{4 \cdot 6} \times \cdots = 2,$$

$$[3] = \frac{1 \cdot 2 \cdot 2 \cdot 2}{4} \times \frac{3 \cdot 3 \cdot 3}{2 \cdot 2 \cdot 5} \times \frac{4 \cdot 4 \cdot 4}{3 \cdot 3 \cdot 6} \times \frac{5 \cdot 5 \cdot 5}{4 \cdot 4 \cdot 7} \times \frac{6 \cdot 6 \cdot 6}{5 \cdot 5 \cdot 8} \times \cdots = 6,$$

and so on, where the infinitude of cancellations serves to obscure questions of convergence. Nonetheless, this infinite product seems to do the trick: if n is a whole number, then $[n] = n!$.

And $[x]$ allows gap-filling. We can consider, for instance, $[1/2]$, which is the value that should be assigned to the interpolation of $(1/2)!$. When Euler substituted $x = 1/2$, he got

$$\left[\frac{1}{2}\right] = \frac{1 \cdot \sqrt{2}}{3/2} \times \frac{\sqrt{2} \cdot \sqrt{3}}{5/2} \times \frac{\sqrt{3} \cdot \sqrt{4}}{7/2} \times \frac{\sqrt{4} \cdot \sqrt{5}}{9/2} \times \cdots$$

$$= \sqrt{\frac{2 \cdot 4}{3 \cdot 3} \times \frac{4 \cdot 6}{5 \cdot 5} \times \frac{6 \cdot 8}{7 \cdot 7} \times \frac{8 \cdot 10}{9 \cdot 9} \times \cdots}.$$

Something about the expression under the radical looked familiar. He recalled a 1655 formula due to John Wallis, who, using an arcane interpolattion procedure of his own, had shown that $\dfrac{3 \cdot 3 \cdot 5 \cdot 5 \cdot 7 \cdot 7 \cdot 9 \cdot 9 \cdot \ldots}{2 \cdot 4 \cdot 4 \cdot 6 \cdot 6 \cdot 8 \cdot 8 \cdot 10 \cdot \ldots} = \dfrac{4}{\pi}$ [22]. With this, Euler deduced that

$$\left[\frac{1}{2}\right] = \sqrt{\frac{\pi}{4}} = \frac{1}{2}\sqrt{\pi}.$$

We are thus forced to conclude that the "natural" interpolation of $\left(\dfrac{1}{2}\right)!$ is the very unnatural $\dfrac{1}{2}\sqrt{\pi}$. That in itself deserves an exclamation point.

This answer provided Euler with a valuable clue. Because π appeared in the result, he surmised that a connection to circular area may lay somewhere beneath the surface, and this, in turn, suggested that he direct his search towards *integrals* [23]. With a bit of effort, he arrived at the alternative formula

$$[x] = \int_0^1 (-\ln t)^x \, dt. \qquad (12)$$

This result is far more compact than (11) and much more elegant. The skeptic can apply equal measures of integration by parts, l'Hospital's rule, and mathematical induction to confirm that, when n is a whole number, $\int_0^1 (-\ln t)^n \, dt = n!$.

Once he had an integral to play with, Euler was in his element. After a few more mathematical gyrations, he found that (see [24])

$$\left[\frac{1}{2}\right] = \sqrt{\int_0^1 \frac{x^2 dx}{\sqrt{1-x^2}} \Bigg/ \int_0^1 \frac{x dx}{\sqrt{1-x^2}}}.$$

A bit of elementary calculus shows that $\int_0^1 \frac{x^2 dx}{\sqrt{1-x^2}} = \frac{\pi}{4}$ and $\int_0^1 \frac{x dx}{\sqrt{1-x^2}} = 1$, so here is another confirmation—this time without resorting to Wallis's formula—that $[1/2] = \sqrt{\frac{\pi}{4}} = \frac{1}{2}\sqrt{\pi}$.

Euler also recognized that $[x] = x \cdot [x-1]$, a relationship he exploited to the hilt in deriving results like $\left[\frac{5}{2}\right] = \frac{5}{2} \times \left[\frac{3}{2}\right] = \frac{5}{2} \times \frac{3}{2} \times \left[\frac{1}{2}\right] = \frac{15}{8}\sqrt{\pi}$ [25]. Then, always a true believer in the persistence of patterns, he pushed the recursion in the other direction to get $\left[\frac{1}{2}\right] = \frac{1}{2} \times \left[-\frac{1}{2}\right]$ and so $\left[-\frac{1}{2}\right] = 2 \times \left[\frac{1}{2}\right] = \sqrt{\pi}$. In other words, $\left(-\frac{1}{2}\right)!$ should be interpreted as $\sqrt{\pi}$. By now it should be evident that intuition has a long way to go to catch up with calculus.

Modern mathematicians tend to follow a modification of Euler's ideas popularized by Adrien-Marie Legendre (1752–1833). Legendre substituted $y = -\ln t$ into (12) to get $[x] = -\int_\infty^0 y^x e^{-y} dy = \int_0^\infty y^x e^{-y} dy$ and then shifted the input by one unit to define the *gamma function* by

$$\Gamma(x) \equiv [x-1] = \int_0^\infty y^{x-1} e^{-y} dy.$$

It is worth noting, however, that this very integral shows up in Euler's writings as well [26].

Of course, the gamma function inherits properties that Euler had discovered about [x], such as the recursion $\Gamma(x + 1) = x\Gamma(x)$ or the remarkable identity $\Gamma(1/2) = [-1/2] = \sqrt{\pi}$. It is a function that seems to appear anywhere sophisticated mathematical analysis is practiced, from probability to differential equations to analytic number theory. Nowadays, the gamma function is regarded as the first and perhaps most important of the "higher functions" of analysis, that is, those whose very definition requires the ideas of calculus. It occupies a place beyond the algebraic, exponential, or trigonometric functions that characterize elementary mathematics. And we owe it, like so much else, to Euler.

The results of this chapter—be they differentials or integrals, approximations or interpolations—reveal an astonishing ingenuity. Von Neumann called Euler "the greatest virtuoso of the period," for he posed the right questions and, with an agility and intuition that continue to amaze, regularly found the right answers [27]. Without doubt, Euler was at home in analysis, the perfect arena in which to apply what seemed to be his informal credo: Follow the formulas, and they will lead to the truth.

No one ever did it better.

CHAPTER 5

First Interlude

Leonhard Euler died in 1783, one year short of the centennial of Leibniz's first paper on differential calculus. By any measure, it had been a remarkable century in the history of mathematics. The results considered thus far, although a tiny fraction of the century's output, illustrate the progress that had been made. Grappling with infinite processes to discover correct and sometimes spectacular results, Newton, Leibniz, the Bernoullis, and Euler had established calculus as the mathematical subdiscipline *par excellence*. Our hats are off to these great originators.

An important trend of that first century was a shift in perspective from the geometric to the analytic. As the problems became more challenging, their solutions depended less on the geometry of curves than on the algebra of functions. The complicated diagrams that Leibniz used to prove his transmutation theorem in 1673 had no counterpart in Euler's work from the middle of the eighteenth century. In this sense, analysis had assumed a more modern look.

But other familiar aspects of the subject were nowhere to be seen. Largely missing, for instance, was that bulwark of modern analysis, the inequality. Seventeenth and eighteenth century mathematicians dealt mainly in *equations*. Their work tended to employ clever substitutions that transformed one formula into another so as to emerge with the desired answer. Although Jakob Bernoulli's divergence proof of the harmonic series (see chapter 3) featured a deft use of inequalities, such an approach was rare.

Rare as well was the study of broad classes of functions. Euler and his predecessors were adept at looking at specific integrals or series, but they were less interested in common properties of, say, continuous or differentiable functions. A shift in focus from the specific to the general would be a hallmark of the coming century.

One other striking difference between early calculus and that of today is the attention given to logical foundations. As we have seen, mathematicians of the period used results whose validity they had neither proved nor, in many cases, even considered. An example was the tendency to replace the integral of an infinite series by the infinite series of integrals, that is, to equate $\int_a^b \left[\sum_{k=1}^{\infty} f_k(x) \right] dx$ and $\sum_{k=1}^{\infty} \left[\int_a^b f_k(x)dx \right]$. Both operations here—integrating functions and summing series—involve infinite processes whose uncritical interchange can lead to incorrect results. Certain conditions must be met before a reversal of this sort is appropriate. On this front, the calculus pioneers operated more on intuition than on reason. Admittedly, their intuition was often very good, with Euler in particular possessing an uncanny ability to know just how far he could go before plunging into the mathematical abyss.

Still, the foundations of calculus were suspect. As an illustration, we recall the role played by infinitely small quantities. Attempts to explain these so-called infinitesimals—and everyone from Leibniz to Euler gave it a shot—never proved satisfactory. Like a mathematical chameleon, infinitesimals seemed inevitably to be both zero and nonzero at the same time. At rock bottom, they were paradoxical, counterintuitive entities.

Nor were things much better when mathematicians based their conclusions on "vanishing" quantities. Newton was a proponent of this dynamic approach, a fitting position, perhaps, for one so captivated by the study of motion. Introducing what we now call the derivative, he considered a quotient of vanishing quantities and wrote that, by the "ultimate ratio" of these evanescent quantities, he meant "the ratio of the quantities not before they vanish, nor afterwards, but with which they vanish" [1]. Besides conjuring up the notion of a quantity *after* it vanishes (whatever that means), Newton asked his readers to imagine a ratio at the precise instant when—poof!—both numerator and denominator simultaneously dissolve into thin air. His description seemed ripe for criticism.

It was not long in coming, and the critic was George Berkeley (1685–1753), noted philosopher and Bishop of Cloyne. In his 1734 essay *The Analyst*, Berkeley ridiculed those scientists who accused him of proceeding on faith and not reason, yet who themselves talked of infinitely small or vanishing quantities. To Berkeley this was at best fuzzy thinking and at worst hypocrisy. The latter was implied in the long subtitle:

A Discourse Addressed to an Infidel Mathematician, wherein It Is Examined Whether the Object, Principles, and Inferences of the Modern Analysis Are More Distinctly Conceived, or More Evidently Deduced, than Religious Mysteries and Points of Faith [2]

Berkeley's essay was caustic. Whether the calculus was built upon Newton's vanishing quantities or Leibniz's infinitely small ones made little difference to the bishop, who concluded that, "The further the mind analyseth and pursueth these fugitive ideas, the more it is lost and bewildered" [3]. When skewering Newton, Berkeley penned the now famous question:

> And what are these fluxions? The velocities of evanescent increments? And what are these same evanescent increments? They are neither finite quantities nor quantities infinitely small, nor yet nothing. May we not call them the ghosts of departed quantities? [4]

He was no kinder to Leibniz's infinitesimals. Admitting that the notion of an infinitely small quantity was "above my capacity," he mockingly observed that an infinitely small part of an infinitely small quantity, for instance, $(dx)^2$, presented "an infinite difficulty to any man whatsoever" [5].

Berkeley did not dispute the conclusions that mathematicians had drawn from these suspect techniques; it was the logic behind them that he rejected. True, the calculus was a wonderful vehicle for finding tangent lines and determining maxima or minima. But he argued that its correct answers arose from incorrect thinking, as certain mistakes cancelled out others in a compensation of errors that obscured the underlying flaws. "Error," he wrote, "may bring forth truth, though it cannot bring forth science" [6].

We illustrate Berkeley's point with his example, using modern notation, of finding $\dfrac{dy}{dx}$ when $y = x^n$. In the fashion of the day, he began by augmenting x with a tiny, nonzero increment o and developing the differential quotient

$$\frac{(x+o)^n - x^n}{o} = \frac{nx^{n-1}o + \dfrac{n(n-1)}{2}x^{n-2}o^2 + \cdots + nxo^{n-1} + o^n}{o}$$

$$= nx^{n-1} + \frac{n(n-1)}{2}x^{n-2}o + \cdots + nxo^{n-2} + o^{n-1}.$$

Up to this point, o was assumed to be nonzero, a supposition, Berkeley stressed, "without which I should not have been able to have made so much as a single step." But then o suddenly became zero, so that

$$\frac{dy}{dx} = nx^{n-1} + 0 + \cdots + 0 = nx^{n-1}.$$

Berkeley objected that the second assumption was in absolute conflict with the first and consequently negated any conclusions derived here. After all, if o is zero, not only are we forbidden to put it into a denominator, but we must concede that x was never augmented at all. The argument collapses in a heap. "When it is said, let the increments vanish," wrote Berkeley, "the former supposition that the increments were something . . . is destroyed, and yet a consequence of that supposition, i.e., an expression got by virtue theoreof, is retained" [7].

To the Bishop, such a method of reasoning was wholly unsatisfactory and represented "a most inconsistent way of arguing, and such as would not be allowed of in Divinity" [8]. In one of *The Analyst's* most searing passages, Berkeley compared the faulty logic of calculus to the high standards that are required "throughout all the branches of humane knowledge, in any other of which, I believe, men would hardly admit such a reasoning as this which, in mathematics, is accepted for demonstration" [9].

Bishop Berkeley had made his point. Although the results of calculus seemed to be valid and, when applied to real-world phenomena like mechanics or optics, yielded solutions that agreed with observations, none of this mattered if the foundations were rotten.

Something had to be done. Over the next decades a number of mathematicians tried to shore up the shaky underpinnings. Among these was Jean-le-Rond d'Alembert (1717–1783), a highly respected scholar who worked alongside Diderot (1713–1784) on the *Encyclopédie* in France. Regarding the foundations of calculus, d'Alembert agreed that infinitely small and/or vanishing quantities were meaningless. He proclaimed, without equivocation, that "a quantity is something or nothing; if it is something, it has not yet vanished; if it is nothing, it has literally vanished. The supposition that there is an intermediate state between these two is a chimera" [10].

As an alternative, d'Alembert proposed that calculus be based upon the concept of *limit*. In treating the derivative, he identified $\frac{dy}{dx}$ as the limit of a quotient of finite terms, which he wrote as $\frac{z}{u}$ but which we recognize as $\frac{y(x + \Delta x) - y(x)}{\Delta x}$. Then, $\frac{dy}{dx}$ is "the quantity to which the ratio z/u approaches more and more closely if we suppose z and u to be real and decreasing. Nothing is clearer than this" [11].

D'Alembert was onto something. He had no use for infinitesimals nor vanishing quantities and deserves credit for highlighting limits as the way to repair the weak foundations of the calculus.

But it would be going too far to assert that d'Alembert saved the day. Although he may have sensed the right path, he did not follow it very far. Missing was a clear definition of "limit" and the subsequent derivation of basic calculus theorems from it. In the end, d'Alembert did little more than suggest the way out of trouble. A full development of these ideas would have to wait a generation and more.

Meanwhile, a greater mathematician weighed in on the matter and offered a very different solution. He was Joseph-Louis Lagrange (1736–1813), a powerful and influential figure in European mathematics as the eighteenth century wound down. On the question of foundations, Lagrange vowed to provide a logically sound framework upon which the great edifice of calculus could be built. In his 1797 work *Théorie des fonctions analytiques*, he envisioned a calculus "freed from all considerations of infinitely small quantities, vanishing quantities, limits and fluxions" [12]. Seeing no merit in any of the past justifications, Lagrange vowed to start anew.

His fundamental idea was to regard infinite series not as the output but as the *source* of differential calculus. That is, beginning with a function $f(x)$ whose derivative he sought, Lagrange expressed $f(x + i)$ as an infinite series in i of the form

$$f(x + i) = f(x) + ip(x) + i^2q(x) + i^3r(x) + \cdots, \tag{1}$$

in which, as he put it, "$p, q, r, \ldots$ will be new functions of x, derived from the primitive function x and independent of the indeterminate i" [13]. Then the (first) derivative of f was no more and no less than $p(x)$, the function serving as the coefficient of i in this expansion.

Anyone familiar with Taylor series can see what Lagrange was up to, but it is important to note that, for him, the series came first and the derivative was a consequence, whereas in modern analysis it is the derivative that precedes the series.

An example might be helpful. Suppose we want to find the derivative $f'(x)$ when $f(x) = \dfrac{1}{x^3}$. (By the way, the "f-prime" notation is due to Lagrange.) Expanding the function as in (1), we have $\dfrac{1}{(x + i)^3} = \dfrac{1}{x^3} + ip(x) + i^2q(x) + i^3r(x) + \cdots$ so that

$$i[p(x) + iq(x) + i^2r(x) + \cdots] = \frac{1}{(x + i)^3} - \frac{1}{x^3} = \frac{-3x^2i - 3xi^2 - i^3}{(x + i)^3x^3}$$

and therefore

$$p(x) + iq(x) + i^2r(x) + \cdots = \frac{-3x^2i - 3xi^2 - i^3}{i(x + i)^3x^3} = \frac{-3x^2 - 3xi - i^2}{(x + i)^3x^3}. \tag{2}$$

At this point, Lagrange let $i = 0$ in (2) to get $p(x) = \dfrac{-3x^2}{x^6}$. Thus, $f'(x) = \dfrac{-3}{x^4}$, which of course would have been no surprise to Newton or Leibniz.

For Lagrange, this derivation avoided quantities that were infinitely small as well as those ghosts of departed quantities vanishing into oblivion. Likewise, he had no need for d'Alembert's uncertainly defined limits. When Lagrange let $i = 0$, he meant that literally. No pitfalls were encountered in (2), for no zero appeared in any denominator. He regarded this as a purely analytic approach to the derivative, one requiring none of the logical gyrations that had embarrassed his predecessors. It was all so neat and tidy.

Or was it? For one thing, defining derivatives in this manner is terribly indirect. The ideas of Newton and Leibniz—even if cluttered with curves and triangles and resting upon a shaky foundation—were at least straightforward in their object. Lagrange's ideas, presented without a single diagram, completely obscured the fact that derivatives had something to do with slopes of tangent lines.

That is a minor criticism. More troubling was the question of how to proceed for less trivial functions than that given above. In our example, the key was to expand and simplify $\dfrac{1}{(x+i)^3} - \dfrac{1}{x^3}$ in order to factor i from the result. But where is the guarantee that every function could be so expanded and simplified? Where is the guarantee that a series so constructed is convergent? And where is the guarantee that a *convergent* series so constructed actually converges to the function we started with? These are deep and important questions.

Ultimately, the theory of Lagrange could not withstand this kind of scrutiny. In 1822 the French mathematician Augustin-Louis Cauchy published an example that proved fatal to Lagrange's ideas. Cauchy, who will be the subject of our next chapter, showed that the function

$$f(x) = \begin{cases} e^{-1/x^2} & \text{if } x \neq 0, \\ 0 & \text{if } x = 0, \end{cases}$$

and all of its derivatives are zero at $x = 0$ [14]. Consequently, as a power series about the origin, $f(x) = 0 + 0 \cdot x + 0 \cdot x^2 + 0 \cdot x^3 + \cdots = 0$, which in turn means that, if we begin with f and write it as a series, we end up with a different function than started with! As a series, we would find it impossible to distinguish between f above and the constant function $g(x) = 0$.

Cauchy's example of two distinct functions sharing a power series indicated that analysis was considerably less benign than Lagrange had assumed.

In the end, a series-based definition of the derivative—and hence a series-based foundation for the calculus—was abandoned. But if Lagrange failed in his primary mission, he made a number of contributions that anticipated the coming century. First, he elevated foundational questions into greater prominence, treating them as both interesting and important issues. Second, he tried to derive the theorems of the calculus from his basic definitions, in the process introducing inequalities and exhibiting skill in their use. Finally, as Judith Grabiner observed in her book, *The Origins of Cauchy's Rigorous Calculus*:

> On reading Lagrange's work, one is struck by his feeling for the general. . . . His extreme love of generality was unusual for this time and contrasts with the emphasis of many of his contemporaries on solving specific problems. His algebraic foundation for the calculus was consistent with his generalizing tendency. [15]

All these contributions notwithstanding, the eighteenth century ended with the logical crisis still unresolved. The work of d'Alembert and Lagrange, along with others who addressed these matters, failed to mollify the critics. As late as 1800, the words of Bishop Berkeley carried the ring of truth: "I say that in every other Science Men prove their Conclusions by their Principles, and not their Principles by the Conclusions" [16].

But a resolution was near. The same Cauchy who recognized the nonuniqueness of series would, in the early nineteenth century, see a way to explain the foundations of calculus in a satisfactory manner. By the time he was done, analysis would be a far more general, abstract, and inequality-laden subject than his predecessors could have imagined. And it would be far more rigorous.

It is to this towering figure, and to his revolution, that we now turn.

CHAPTER 6

Cauchy

Augustin-Louis Cauchy

Eric Temple Bell, who popularized mathematicians in colorful if sometimes immoderate prose, wrote that "Cauchy's part in modern mathematics is not far from the center of the stage" [1]. It is hard to argue with this judgment. During his career, Augustin-Louis Cauchy (1789–1857) published books and papers that now fill over two dozen volumes of collected works, and among these are treatises on combinatorics and algebra, differential equations and complex variables, mechanics, and optics. Like Leonhard Euler from the century before, Augustin-Louis Cauchy cast a long shadow.

His impact upon the history of calculus is especially profound. Cauchy stands at a boundary between the early practitioners, who, for all their cleverness, occupied a more intuitive, more innocent world, and the mathematicians of today, for whom the logical standards are strict, pervasive, and unforgiving. Cauchy did not complete this transformation, for his ideas would require considerable fine tuning in the decades to come. But the similarity between Cauchy's development of analysis and that of today's textbooks cannot fail to impress the modern reader.

This chapter gives a taste of Cauchy in action. We include a number of examples, ranging from his theory of limits to the mean value theorem and from his definition of the integral to the fundamental theorem of calculus, before concluding with a pair of tests for series convergence. This material comes from two great texts: his 1821 *Cours d'analyse de l'École Royale Polytechnique* and his 1823 *Résumé des leçons données a l'École Royale Polytechnique, sur le calcul infinitésimal* [2].

LIMITS, CONTINUITY, AND DERIVATIVES

Although Cauchy recognized Lagrange as an elder statesman of mathematics, he could not endorse the latter's series-based definition of the derivative. "I reject the development of functions by infinite series," wrote Cauchy, who continued:

> I do not ignore that the illustrious [Lagrange] has taken this formula as the basis for his theory of derived functions. But, in spite of the respect commanded by so great an authority, most geometers now acknowledge the uncertainty of results to which one can be led by use of divergent series . . . and we add that [Lagrange's methods] lead to the development of a function by a convergent series, although the sum of this series differs essentially from the function proposed. [3]

The last allusion is to Cauchy's counterexample mentioned in the previous chapter. For him, Lagrange's program was a dead end. Hoping to provide a logically valid alternative, Cauchy asserted that "the principles of differential calculus, and their most important applications, can easily be developed without the need of series."

Instead, Cauchy believed that the foundation upon which all calculus would be built was the idea of *limit*. His definition of this concept is a mathematical classic:

> When the values successively attributed to a variable approach indefinitely to a fixed value, in a manner so as to end by differing from it by as little as one wishes, this last is called the limit of all the others. [4]

Cauchy gave the example of a circle's area as the limit of the areas of inscribed regular polygons as the number of sides increases without bound. Of course, no polygonal area ever *equals* that of the circle. But for any proposed tolerance, an inscribed regular polygon can be found whose area, and those of all inscribed regular polygons with even more sides, is closer to that of the circle than the tolerance stipulated. Polygonal areas get close—and stay close—to the area of the circle. This is the essence of Cauchy's idea.

A modern reader may be surprised by his definition's wordiness, its dynamic imagery, and the absence of ε and δ. Nowadays we do not talk about a "succession" of numbers "approaching" something, and we tend to prefer the symbolic efficiency of "$\varepsilon > 0$" to the phrase "as little as one wishes."

Yet this was an advance of the first order. Cauchy's idea, based on "closeness," avoided some of the pitfalls of earlier attempts. In particular, he said nothing about reaching the limit nor about surpassing it. Such issues ensnared many of Cauchy's predecessors, as Berkeley had been only too happy to point out. By contrast, Cauchy's so-called "limit avoidance" definition made no mention whatever of attaining the limit, just of getting and staying close to it. For him, there *were* no departed quantities, and Berkeley's ghosts disappeared.

Cauchy introduced a related concept that may raise a few eyebrows. He wrote that "when the successive numerical values of a variable decrease indefinitely (so as to become less than any given number), this variable will be called . . . an infinitely small quantity" [5]. His use of "infinitely small" strikes us as unfortunate, but we can regard this definition as simply spelling out what is meant by convergence to zero.

Cauchy next turned his attention to continuity. Intuition might at first suggest that he had things backwards, that he should have based the idea of limits upon that of continuity and not vice versa. But Cauchy had it right. Reversing the "obvious" order of affairs was the key to understanding continuous functions.

Starting with $y = f(x)$, he let i be an infinitely small quantity (as defined above) and considered the function's value when x was replaced by $x + i$. This changed the functional value from y to $y + \Delta y$, a relationship Cauchy expressed as

$$y + \Delta y = f(x + i) \quad \text{or} \quad \Delta y = f(x + i) - f(x).$$

If, for i infinitely small, the difference $\Delta y = f(x + i) - f(x)$ was infinitely small as well, Cauchy called f a *continuous* function of x [6]. In other words, a function is continuous at x if, when the independent variable x is augmented by an infinitely small quantity, the dependent variable y likewise grows by an infinitely small amount.

Again, reference to the "infinitely small" means only that the quantities have limit zero. In this light, we see that Cauchy has called f continuous at x if $\lim_{i \to 0}[f(x + i) - f(x)] = 0$, which is equivalent to the modern definition, $\lim_{i \to 0} f(x + i) = f(x)$.

As an illustration, Cauchy considered $y = \sin x$ [7]. He used the fact that $\lim_{x \to 0}(\sin x) = 0$ as well as the trig identity $\sin(\alpha + \beta) - \sin\alpha = 2\,\sin(\beta/2) \cdot \cos(\alpha + \beta/2)$. Then, for infinitely small i, he observed:

$$\Delta y = f(x + i) - f(x) = \sin(x + i) - \sin x = 2\sin(i/2)\cos(x + i/2). \quad (1)$$

Because $i/2$ is infinitely small, so is $\sin(i/2)$ and so too is the entire right-hand side of (1). By Cauchy's definition, the sine function is continuous at any x.

We note that Cauchy also recognized one of the most important properties of continuous functions: their preservation of sequential limits. That is, if f is continuous at a and if $\{x_k\}$ is a sequence for which $\lim_{k \to \infty} x_k = a$, then it follows that $\lim_{k \to \infty} f(x_k) = f\left[\lim_{k \to \infty} x_k\right] = f(a)$. We shall see him exploit this principle shortly.

He then considered "derived functions." For Cauchy, the differential quotient was defined as

$$\frac{\Delta y}{\Delta x} = \frac{f(x + i) - f(x)}{i},$$

where i is infinitely small. Taking his notation from Lagrange, Cauchy denoted the derivative by y' or $f'(x)$ and claimed that this was "easy" to determine for simple functions like

$$y = r \pm x,\ rx,\ r/x,\ x^r,\ A^x,\ \log_A x,\ \sin x,\ \cos x,\ \arcsin x, \text{ and } \arccos x.$$

We shall examine just one of these: $y = \log_A x$, the logarithm to base $A > 1$, which Cauchy denoted by $L(x)$ [8].

He began with the differential quotient $\dfrac{\Delta y}{\Delta x} = \dfrac{f(x + i) - f(x)}{i} = \dfrac{L(x + i) - L(x)}{i}$ for i infinitely small and introduced the auxiliary variable

$\alpha = \dfrac{i}{x}$, which is infinitely small as well. Using rules of logarithms and substituting liberally, Cauchy reasoned that

$$\frac{\Delta y}{\Delta x} = \frac{L(x+i) - L(x)}{i} = \frac{L\left(\dfrac{x+i}{x}\right)}{i} = \frac{L\left(\dfrac{x+\alpha x}{x}\right)}{\alpha x}$$

$$= \frac{\dfrac{1}{\alpha}L(1+\alpha)}{x} = \frac{1}{x}L(1+\alpha)^{1/\alpha}. \tag{2}$$

For α infinitely small, he identified this last expression as $\dfrac{1}{x}L(e)$. Today we would invoke continuity of the logarithm and the fact that $\lim\limits_{\alpha \to 0}(1+\alpha)^{1/\alpha} = e$ to justify this step. In any case, Cauchy concluded from (2) that the derivative of $L(x)$ was $\dfrac{1}{x}L(e)$. As a corollary, he noted that the derivative of the *natural* logarithm $\ln(x)$ is $\dfrac{1}{x}\ln(e) = \dfrac{1}{x}$.

He obviously had his differential calculus well under control.

THE INTERMEDIATE VALUE THEOREM

Cauchy's analytic reputation rests not only upon his definition of the limit. At least as significant was his recognition that the great theorems of calculus must be proved from this definition. Whereas earlier mathematicians had accepted certain results as true because they either conformed to intuition or were supported by a diagram, Cauchy seemed unsatisfied unless an algebraic argument could be advanced to prove them. He left no doubt of his position when he wrote that "it would be a serious error to think that one can find certainty only in geometrical demonstrations or in the testimony of the senses" [9].

His philosophy was evident in a demonstration of the intermediate value theorem. This famous result begins with a function f continuous between x_0 and X (Cauchy's preferred designation for the endpoints of an interval). If $f(x_0) < 0$ and $f(X) > 0$, the intermediate value theorem asserts that the function *must* equal zero at one or more points between x_0 and X.

For those who trust their eyes, nothing could be more obvious. An object moving continuously from a negative to a positive value must *somewhere* slice across the x-axis. As indicated in figure 6.1, the intermediate value occurs at $x = a$, where $f(a) = 0$. It is tempting to ask, "What's the big deal?"

Of course, the big deal is that mathematicians hoped to free analysis from the danger of intuition and the allure of geometry. For Cauchy, even obvious things had to be proved with indisputable logic.

In that spirit, he began his proof of the intermediate value theorem by letting $h = X - x_0$ and fixing a whole number $m > 1$ [10]. He then broke the interval from x_0 to X into m equal subintervals at the points x_0, $x_0 + h/m$, $x_0 + 2h/m$, ..., $X - h/m$, X and considered the related sequence of functional values:

$$f(x_0), f(x_0 + h/m), f(x_0 + 2h/m), \ldots, f(X - h/m), f(X).$$

Because the first of these was negative and the last positive, he observed that, as we progress from left to right, we will find two consecutive functional values with opposite signs. More precisely, for some whole number n, we have

$$f(x_0 + nh/m) \le 0 \quad \text{but} \quad f(x_0 + (n+1)h/m) \ge 0.$$

We follow Cauchy in denoting these consecutive points of subdivision by $x_0 + nh/m \equiv x_1$ and $x_0 + (n+1)h/m \equiv X_1$. Clearly, $x_0 \le x_1 < X_1 \le X$, and the length of the interval from x_1 to X_1 is h/m.

He now repeated the procedure across the smaller interval from x_1 to X_1. That is, he divided it into m equal subintervals, each of length h/m^2, and considered the sequence of functional values

$$f(x_1), f(x_1 + h/m^2), f(x_1 + 2h/m^2), \ldots, f(X_1 - h/m^2), f(X_1).$$

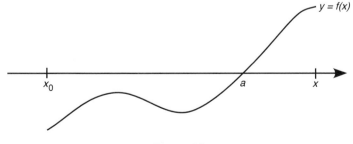

Figure 6.1

Again, the leftmost value is less than or equal to zero, whereas the rightmost is greater than or equal to zero, so there must be consecutive points x_2 and X_2 a distance of h/m^2 units apart, for which $f(x_2) \le 0$ and $f(X_2) \ge 0$. At this stage, we have $x_0 \le x_1 \le x_2 < X_2 \le X_1 \le X$. Those familiar with the bisection method for approximating solutions to equations should feel perfectly at home with Cauchy's procedure.

Continuing in this manner, he generated a nondecreasing sequence $x_0 \le x_1 \le x_2 \le x_3 \le \cdots$ and a nonincreasing sequence $\cdots \le X_3 \le X_2 \le X_1 \le X$, where all the values $f(x_k) \le 0$ and $f(X_k) \ge 0$ and for which the gap $X_k - x_k = h/m^k$. For increasing k, this gap obviously decreases toward zero, and from this Cauchy concluded that the ascending and descending sequences must converge to a common limit a. In other words, there is a point a for which $\lim_{k \to \infty} x_k = a = \lim_{k \to \infty} X_k$.

We pause to comment on this last step. Cauchy here assumed a version of what we now call the completeness property of the real numbers. He took it for granted that, because the terms of the sequences $\{x_k\}$ and $\{X_k\}$ grow arbitrarily close to one another, they must converge to a common limit. One could argue that his belief in the existence of this point a is as much a result of unexamined intuition as simply believing the intermediate value theorem in the first place. But such a judgment may be overly harsh. Even if Cauchy invoked an untested hypothesis, he had at least pushed the argument much deeper toward the core principles. If he failed to clear the path of all obstacles, he got rid of most of the brush underfoot.

To finish the argument, Cauchy stated (without proof) that the point a falls within the original interval from x_0 to X, and then he used the continuity of f to conclude, in modern notation, that

$$f(a) = f\left[\lim_{k \to \infty} x_k\right] = \lim_{k \to \infty} f(x_k) \le 0 \quad \text{and}$$

$$f(a) = f\left[\lim_{k \to \infty} X_k\right] = \lim_{k \to \infty} f(X_k) \ge 0.$$

In Cauchy's words, these inequalities established that "the quantity $f(a) \ldots$ cannot differ from zero." He had thus proved the *existence* of a number a between x and X for which $f(a) = 0$. The general version of the intermediate value theorem, namely that a continuous function takes *all* values between $f(x_0)$ and $f(X)$, follows as an easy corollary.

This was a remarkable achievement. Cauchy had, for the most part, succeeded in demonstrating a "self-evident" principle by analytic methods.

As Judith Grabiner observed, "though the mechanics of the proof are simple, the basic conception of the proof is revolutionary. Cauchy transformed the approximation technique into something entirely different: a proof of the existence of a limit" [11].

THE MEAN VALUE THEOREM

We now turn to another staple of the calculus, the mean value theorem for derivatives [12]. In his *Calcul infinitésimal*, Cauchy began with a preliminary result.

Lemma: If, for a function f continuous between x_0 and X, one lets A be the smallest and B be the largest value that f' takes on this interval, then

$$A \leq \frac{f(X) - f(x_0)}{X - x_0} \leq B.$$

Proof: We note that Cauchy's reference to f'—and thus his unstated assumption that f is differentiable—would of course guarantee the continuity of f. Moreover, he assumed outright that the derivative takes a greatest and least value on the interval $[x_0, X]$. A modern approach would treat these hypotheses with more care.

If his statement seems peculiar, his proof began with a now-familiar ring, for Cauchy introduced two "very small numbers" δ and ε. These were chosen so that, for all positive values of $i < \delta$ and for any x between x_0 and X, we have

$$f'(x) - \varepsilon < \frac{f(x + i) - f(x)}{i} < f'(x) + \varepsilon. \tag{3}$$

Here Cauchy was assuming a uniformity condition for his choice of δ. The existence of the derivative certainly means that, for any $\varepsilon > 0$ and for any fixed x, there is a $\delta > 0$ for which the inequalities of (3) hold. But such a δ depends on both ε and the particular point x. Without additional results or assumptions, Cauchy could not justify the choice of a single δ that simultaneously works for all x throughout the interval.

Be that as it may, he next subdivided the interval by choosing points

$$x_0 < x_1 < x_2 < \cdots < x_{n-1} < X,$$

where $x_1 - x_0$, $x_2 - x_1$, ... , $X - x_{n-1}$ "have numerical values less than δ." For these subdivisions, repeated applications of (3) and the fact that $A \leq f'(x) \leq B$ imply that

$$A - \varepsilon < f'(x_0) - \varepsilon < \frac{f(x_1) - f(x_0)}{x_1 - x_0} < f'(x_0) + \varepsilon < B + \varepsilon,$$

$$A - \varepsilon < f'(x_1) - \varepsilon < \frac{f(x_2) - f(x_1)}{x_2 - x_1} < f'(x_1) + \varepsilon < B + \varepsilon,$$

$$\cdot$$
$$\cdot$$
$$\cdot$$

$$A - \varepsilon < f'(x_{n-1}) - \varepsilon < \frac{f(X) - f(x_{n-1})}{X - x_{n-1}} < f'(x_{n-1}) + \varepsilon < B + \varepsilon.$$

Cauchy then observed that, "if one divides the sum of these numerators by the sum of these denominators, one obtains a *mean* fraction which is . . . contained between the limits $A - \varepsilon$ and $B + \varepsilon$." Here he was using the fact that, if $b_k > 0$ for $k = 1, 2, \ldots, n$ and if $C < \dfrac{a_k}{b_k} < D$ for all k, then $C < \displaystyle\sum_{k=1}^{n} a_k \Big/ \sum_{k=1}^{n} b_k < D$ as well. Applying this result to the inequalities above, he found that

$$A - \varepsilon < \frac{f(x_1) - f(x_0) + f(x_2) - f(x_1) + \cdots + f(X) - f(x_{n-1})}{(x_1 - x_0) + (x_2 - x_1) + \cdots + (X - x_{n-1})} < B + \varepsilon,$$

which telescoped to $A - \varepsilon < \dfrac{f(X) - f(x_0)}{X - x_0} < B + \varepsilon.$ Cauchy ended the proof with the statement that, "as this conclusion holds however small be the number ε, one can affirm that the expression $\left[\dfrac{f(X) - f(x_0)}{X - x_0} \right]$ will be bounded between A and B."   Q.E.D.

This is an interesting argument, one that stumbles over the issue of uniformity yet demonstrates a genius in working with inequalities and employing the now-ubiquitous ε and δ to reach its desired conclusion. No one would confuse this level of generality and rigor with something from the early days of Newton and Leibniz.

Cauchy then used the lemma to prove his mean value theorem.

Theorem: If the function f and its derivative f' are continuous between x_0 and X, then for some θ between 0 and 1, we have

$$\frac{f(X) - f(x_0)}{X - x_0} = f'[x_0 + \theta(X - x_0)].$$

Proof: The assumed continuity of f' guarantees, by the general version of the intermediate value theorem, that f' must take any value between its least (A) and its greatest (B). But according to the lemma, the number $\frac{f(X) - f(x_0)}{X - x_0}$ is one such intermediate value, and so, as Cauchy put it, "there exists between the limits 0 and 1 a value of θ sufficient to satisfy the equation

$$\frac{f(X) - f(x_0)}{X - x_0} = f'[x_0 + \theta(X - x_0)]." \tag{4}$$

Q.E.D.

The conclusion in (4) differs from what we find in a modern textbook only in the notational convention that replaces Cauchy's $x_0 + \theta(X - x_0)$ by our c, where of course $0 < \theta < 1$ implies $x_0 < c < X$.

So, this is the mean value theorem for derivatives, albeit proved under Cauchy's assumption that the derivative is continuous, an assumption made to guarantee that f' takes all intermediate values between A and B. In fact, this assumption is unnecessary, and modern proofs of the mean value theorem get along quite nicely without it. Moreover, it turns out that derivatives take intermediate values whether or not they are continuous, a striking result we shall prove in chapter 10.

In the 1820s, these finer points were unclear, and Cauchy's insight, significant for its time, would not be the final word. Nevertheless, he had identified the mean value theorem as central to a rigorous development of the calculus, a position it retains to this day.

INTEGRALS AND THE FUNDAMENTAL THEOREM OF CALCULUS

Like Cauchy's approach to limits, his definition of the integral would reverberate through the history of calculus. We recall that Leibniz had defined the integral as a sum of infinitely many infinitesimal summands and chose the notation $\int$ to suggest this. Strange as it may seem, by 1800 integration was no longer perceived in this light. Rather, it had come to be regarded primarily as the inverse of differentiation, occupying a secondary position in the pantheon of mathematical concepts. Euler, for instance, began his influential three-volume text on *integral* calculus with the following:

> **Definition:** Integral calculus is the method of finding, from a given differential, the quantity itself; and the operation which produces this is generally called integration. [13]

Euler thought of integration as dependent upon, and hence subservient to, differentiation.

Cauchy disagreed. He believed the integral must have an independent existence and defined it accordingly. He thereby initiated a transformation that, as the nineteenth century wore on, would catapult integration into the analytic spotlight.

He began with a function f continuous on the interval between x_0 and X [14]. Although continuity was critical to his definition, Cauchy pointedly did not assume that f was the derivative of some other function. He subdivided the interval into what he called "elements" $x_1 - x_0$, $x_2 - x_1, x_3 - x_2, \ldots, X - x_{n-1}$ and let

$$S = (x_1 - x_0)f(x_0) + (x_2 - x_1)f(x_1) + (x_3 - x_2)f(x_2)$$
$$+ \cdots + (X - x_{n-1})f(x_{n-1}).$$

We recognize this as a sum of left-hand rectangular areas, but in his *Calcul infinitésimal*, Cauchy made no mention of the geometry of the situation nor did he provide the now-customary diagram. He did, however, observe that "the quantity S clearly depends on: (1) the number n of elements into which we have divided the difference $X - x_0$; (2) the values of these elements and, as a consequence, the mode of division adopted." Further, he claimed that "it is important to note that, if the numerical values of the elements differ very little and the number n is quite large, then the manner of division will have an imperceptible effect on the value of S."

Cauchy gave an argument in support of this last assertion, one that assumed uniform continuity—"one δ fits all"—without recognizing it. In this way, he believed he had proved the following result:

> If we decrease indefinitely the numerical values of these elements [that is, of $x_1 - x_0, x_2 - x_1, x_3 - x_2, \ldots, X - x_{n-1}$] while augmenting their number, the value of S ... ends by attaining a certain limit that depends uniquely on the form of the function $f(x)$ and

the extreme values x_0 and X attained by the variable x. This limit is what we call a definite integral.

He followed Joseph Fourier (1768–1830) in adopting $\int_{x_0}^{X} f(x)dx$ as "the most simple" notation for the limit in question.

Cauchy's definition was far from perfect, in large measure because it applied only to continuous functions. Still, it was a highly significant development that left no doubt about two critical points: (1) the integral was a *limit* and (2) its existence had nothing to do with antidifferentiation.

As was his custom, Cauchy used the definition to prove basic results. Some were general rules, such as the fact that the integral of the sum is the sum of the integrals. Others were specific formulas like

$$\int_{x_0}^{X} x\,dx = \frac{X^2 - x_0^2}{2} \quad \text{or} \quad \int_{x_0}^{X} \frac{dx}{x} = \ln\left(\frac{X}{x_0}\right).$$ And Cauchy established that,

for f continuous, there exists a value of θ between 0 and 1 for which

$$\int_{x_0}^{X} f(x)dx = (X - x_0)f[x_0 + \theta(X - x_0)]. \tag{5}$$

Readers will recognize this as the mean value theorem for integrals.

Only then, having come this far without even mentioning derivatives, was Cauchy ready to bind together the great ideas of differentiation and integration. The unifying result is what we call the fundamental theorem of calculus. As one of the great theorems in all of mathematics, proved by one of the great analysts of all time, it surely deserves our attention [15].

As usual, Cauchy began with a continuous function f, but this time, in considering its integral, he let the upper limit of integration vary. That is,

he defined the function $\Phi(x) = \int_{x_0}^{x} f(x)dx$, although in the interest of

clarity we now would write $\Phi(x) = \int_{x_0}^{x} f(t)dt$. Cauchy argued that

$$\Phi(x + \alpha) - \Phi(x) = \int_{x_0}^{x+\alpha} f(x)dx - \int_{x_0}^{x} f(x)dx$$

$$= \int_{x_0}^{x} f(x)dx + \int_{x}^{x+\alpha} f(x)dx - \int_{x_0}^{x} f(x)dx$$

$$= \int_{x}^{x+\alpha} f(x)dx.$$

Moreover, by (5), there exists θ between 0 and 1 for which

$$\int_{x}^{x+\alpha} f(x)dx = (x + \alpha - x)f[x + \theta(x + \alpha - x)] = \alpha f(x + \theta\alpha).$$

In short, $\Phi(x + \alpha) - \Phi(x) = \alpha f(x + \theta\alpha)$ for some value of θ.

To Cauchy, this last equation showed that Φ was continuous because an infinitely small increase in x produces an infinitely small increase in Φ. Or, as we might put it,

$$\lim_{\alpha \to 0}[\Phi(x + \alpha) - \Phi(x)] = \lim_{\alpha \to 0} \alpha f(x + \theta\alpha) = \lim_{\alpha \to 0} \alpha \cdot \lim_{\alpha \to 0} f(x + \theta\alpha)$$

$$= \lim_{\alpha \to 0} \alpha \cdot f(\lim_{\alpha \to 0}[x + \theta\alpha]) = 0 \cdot f(x) = 0,$$

where the continuity of f at x implies $\lim_{\alpha \to 0} f(x + \theta\alpha) = f(x)$. Consequently,

$\lim_{\alpha \to 0} \Phi(x + \alpha) = \Phi(x)$ and so Φ is continuous at x.

But Cauchy was after bigger game, for it also followed that

$$\Phi'(x) = \lim_{\alpha \to 0}\left[\frac{\Phi(x + \alpha) - \Phi(x)}{\alpha}\right] = \lim_{\alpha \to 0}\frac{\alpha f(x + \theta\alpha)}{\alpha}$$

$$= \lim_{\alpha \to 0} f(x + \theta\alpha) = f(x).$$

Just to be sure no one missed the point, Cauchy rephrased this as

$$\frac{d}{dx}\int_{x_0}^{x} f(x)dx = f(x). \tag{6}$$

This is the "first version" of the fundamental theorem of calculus. In equation (6), the inverse nature of differentiation and integration jumps right off the page.

Having differentiated the integral, Cauchy next showed how to integrate the derivative. He began with a simple but important result that he called a "problem."

Problem: If ω is a function whose derivative is everywhere zero, then ω is constant.

Proof: We fix x_0 in the function's domain. If x is another point in the domain, the mean value theorem (4) guarantees a θ between 0 and 1 such that

$$\frac{\omega(x) - \omega(x_0)}{x - x_0} = \omega'[x_0 + \theta(x - x_0)] = 0,$$

and so $\omega(x) = \omega(x_0)$. Cauchy continued, "If one designates by c the constant quantity $\omega(x_0)$, then $\omega(x) = c$" for all x. In short, ω is constant as required.   Q.E.D.

He was now ready for the second version of the fundamental theorem. Cauchy assumed that f is continuous and that F is a function with $F'(x) = f(x)$ for all x. If $\Phi(x) = \int_{x_0}^{x} f(x)dx$, he knew from (6) that $\Phi'(x) = f(x)$. Letting $\omega(x) = \Phi(x) - F(x)$, Cauchy reasoned that

$$\omega'(x) = \Phi'(x) - F'(x) = f(x) - f(x) = 0.$$

Thus there is a constant c with $c = \omega(x) = \Phi(x) - F(x)$. He substituted $x = x_0$ into this last equation to get

$$c = \Phi(x_0) - F(x_0) = \int_{x_0}^{x_0} f(x)dx - F(x_0) = 0 - F(x_0) = -F(x_0).$$

It follows that $\int_{x_0}^{x} f(x)dx = \Phi(x) = F(x) + c = F(x) - F(x_0)$. After changing the upper limit of integration to X, Cauchy had what he wanted:

$$\int_{x_0}^{X} f(x)dx = F(X) - F(x_0). \tag{7}$$

---

(16) $$\tilde{\mathfrak{F}}(x) = \int_{x_0}^{x} f(x)\,dx = \mathbf{F}(x) + \varpi(x).$$

Si, de plus, les fonctions $f(x)$ et $\mathbf{F}(x)$ sont l'une et l'autre continues entre les limites $x = x_0$, $x = \mathbf{X}$, la fonction $\tilde{\mathfrak{F}}(x)$ sera elle-même continue, et par suite $\varpi(x) = \tilde{\mathfrak{F}}(x) - \mathbf{F}(x)$ conservera constamment la même valeur entre ces limites, entre lesquelles on aura

$$\varpi(x) = \varpi(x_0),$$

$$\tilde{\mathfrak{F}}(x) - \mathbf{F}(x) = \tilde{\mathfrak{F}}(x_0) - \mathbf{F}(x_0) = -\mathbf{F}(x_0), \qquad \tilde{\mathfrak{F}}(x) = \mathbf{F}(x) - \mathbf{F}(x_0),$$

(17) $$\int_{x_0}^{x} f(x)\,dx = \mathbf{F}(x) - \mathbf{F}(x_0).$$

Enfin, si dans l'équation (17) on pose $x = \mathbf{X}$, on trouvera

(18) $$\int_{x_0}^{X} f(x)\,dx = \mathbf{F}(\mathbf{X}) - \mathbf{F}(x_0).$$

Cauchy's proof of the fundamental theorem of calculus (1823)

To see the inverse relationship, we need only replace $f(x)$ by $F'(x)$ and write (7) as $\int_{x_0}^{X} F'(x)dx = F(X) - F(x_0)$. This version of the fundamental theorem integrates the derivative, thereby complementing its predecessor.

So, when integrating a continuous function f across the interval from x_0 to X, we can short-circuit Cauchy's intricate definition with its "elements" and sums and limits *provided* we find an antiderivative F. In this happy circumstance, evaluating the integral becomes nothing more than substituting x_0 and X into F. One could argue that (7) represents the greatest shortcut in all of mathematics.

Although the fundamental theorem is a fitting capstone to any rigorous development of calculus, we end this chapter in yet another corner of analysis where Cauchy made a significant impact: the realm of infinite series.

TWO CONVERGENCE TESTS

Like Newton, Leibniz, and Euler before him, Cauchy was a master of infinite series. But unlike these predecessors, he recognized the need to treat questions of convergence/divergence with care, lest divergent series lead mathematicians astray. If Cauchy held such a position, it seemed incumbent upon him to supply tests for convergence, and on this front he did not disappoint.

First we must say a word about Cauchy's definition of the sum of an infinite series. Earlier mathematicians, who could be amazingly clever in evaluating specific series, tended to treat these holistically, as single expressions that behaved more or less like their finite counterparts. To Cauchy, the meaning of $\sum_{k=0}^{\infty} u_k$ was more subtle. It required a precise definition in order to determine not only its value but its very *existence*.

His approach is now familiar. Cauchy introduced the sequence of partial sums

$$S_1 = u_0,\; S_2 = u_0 + u_1,\; S_3 = u_0 + u_1 + u_2, \text{ and generally } S_n = \sum_{k=0}^{n-1} u_k.$$

Then the value of the infinite series was defined to be the limit of this sequence, that is, $\sum_{k=0}^{\infty} u_k \equiv \lim_{n\to\infty} S_n = \lim_{n\to\infty} \sum_{k=0}^{n-1} u_k$, provided the limit exists, in which case "the series will be called *convergent* and the limit . . . will be called the sum of the series" [16]. As he had done with derivatives and integrals, Cauchy erected a theory of infinite series upon the bedrock of limits.

It was an ingenious idea, although in the process Cauchy committed an error of omission. From time to time, he asserted the existence of the limit of a sequence of partial sums based on the fact that the partial sums grew ever closer to one another. By this last statement he meant that, for any $\varepsilon > 0$, there is an index N so that the difference between S_N and S_{N+k} is less than ε for all $k \geq 1$. In his honor, we now call a sequence with this property a "Cauchy sequence."

However, he offered no justification for the idea that terms growing arbitrarily close to one another must necessarily converge to some limit. As noted above, this condition is an alternative version of the completeness property, the logical foundation upon which the theory of limits, and hence the theory of calculus, now rests. To modern mathematicians, completeness must be addressed either by deriving it from a more elementary definition of the real numbers or by adopting it as an axiom. One could argue that Cauchy more or less did the latter, although there is a difference between assuming something explicitly (as an axiom) and assuming it implicitly (as a gaffe).

In any case, he treated as self-evident the fact that a Cauchy sequence is convergent. There is an irony here, for we now attach his name to a concept he did not fully comprehend. But rather than diminish his status, this irony reinforces our previous observation that difficult ideas take time to reach maturity.

With that prologue, we now consider a pair of tests with which Cauchy demonstrated the convergence of infinite series. Both proofs are based on the comparison test for a series of nonnegative terms, which says that if $0 \leq a_k \leq b_k$ for all k and if $\displaystyle\sum_{k=0}^{\infty} b_k$ converges, then so does $\displaystyle\sum_{k=0}^{\infty} a_k$. Today the comparison test is proved by means of the aforementioned completeness property, and it remains one of the easiest ways to establish series convergence.

The first of our results, the root test, he stated in the following words.

Theorem: For the infinite series $u_0 + u_1 + u_2 + u_3 + \cdots + u_k + \cdots$, find the limit or limits to which the expression $|u_k|^{1/k} = \sqrt[k]{|u_k|}$ converges and let λ be the greatest of these. Then the series converges if $\lambda < 1$ and diverges if $\lambda > 1$ [17].

Before proceeding, we should clarify a few points. For one, Cauchy did not use the absolute value notation, as we have. Rather, he talked about ρ_k as the "numerical value" or the "modulus" of u_k and framed the root test in terms of ρ_k. Of course, this is just a symbolic convention, not a substantive difference.

Perhaps less familiar is his reference to the λ as the "greatest" of the limits. Again, we now have a term for this, the *limit supremum*, and we write $\lambda = \limsup |u_k|^{1/k}$ or $\lambda = \overline{\lim} |u_k|^{1/k}$ in place of Cauchy's verbal description.

For readers unfamiliar with the concept, an example may be useful. Suppose we consider the infinite series $\sum_{k=0}^{\infty} u_k = 1 + \dfrac{1}{3} + \dfrac{1}{4} + \dfrac{1}{27} + \dfrac{1}{16} + \dfrac{1}{243} + \dfrac{1}{64} + \dfrac{1}{2187} + \cdots$, where reciprocals of certain powers of 3 alternate with those of certain powers of 2. We see that the series terms $u_0, u_1, u_2, u_3, \ldots$ obey the pattern:

$$
\begin{cases}
u_{2k} = \dfrac{1}{2^{2k}} & \text{for } k = 0, 1, 2, \ldots, \\[2ex]
u_{2k+1} = \dfrac{1}{3^{2k+1}} & \text{for } k = 0, 1, 2, \ldots.
\end{cases}
$$

If we look only at terms with even subscripts, we find the limit of their roots to be $\lim_{k\to\infty} \sqrt[2k]{1/2^{2k}} = \dfrac{1}{2}$, whereas if we restrict ourselves to terms with odd subscripts, we have $\lim_{k\to\infty} \sqrt[2k+1]{1/3^{2k+1}} = \dfrac{1}{3}$. In modern parlance, the sequence $\{|u_k|^{1/k}\}$ has a subsequence converging to $\dfrac{1}{2}$ and another converging to $\dfrac{1}{3}$. In this case, the greater is $\lambda = \dfrac{1}{2}$.

Cauchy's proof of the root test in *Calcul infinitésimal* is virtually identical to that found in a modern text. He began with the case where $0 < \lambda < 1$ and fixed a number μ so that $\lambda < \mu < 1$. His critical observation was that the "greatest values" of $|u_k|^{1/k}$ "cannot approach indefinitely the limit λ without eventually becoming less than μ." As a consequence, he knew there was an integer m such that, for all $k \geq m$, we have $|u_k|^{1/k} < \mu$ and so $|u_k| < \mu^k$. He then considered the two infinite series

$$|u_m| + |u_{m+1}| + |u_{m+2}| + \cdots \leq \mu^m + \mu^{m+1} + \mu^{m+2} + \cdots,$$

where the geometric series on the right converges because $\mu < 1$. From the comparison test, Cauchy deduced the convergence of $\sum_{k=0}^{\infty} |u_k|$, and thus of $\sum_{k=0}^{\infty} u_k$ as well. In short, if $\lambda < 1$, the series converges. It follows, for instance, that the series $1 + \dfrac{1}{3} + \dfrac{1}{4} + \dfrac{1}{27} + \dfrac{1}{16} + \dfrac{1}{243} + \dfrac{1}{64} + \dfrac{1}{2187} + \cdots$ converges because $\lambda = 1/2$.

His proof of the divergence case ($\lambda > 1$) was analogous. To demonstrate the importance of the root test, Cauchy applied it to determine what we now call the radius of convergence of the Maclaurin series $\sum_{k=0}^{\infty} \dfrac{f^{(k)}(0)}{k!} x^k$, and from there a rigorous theory of power series was on its way.

There are other tests of convergence scattered through Cauchy's collected works, such as the ratio test (credited to d'Alembert) and the Cauchy condensation test [18]. The latter begins with a series $\sum_{k=0}^{\infty} u_k$, where $u_0 \geq u_1 \geq u_2 \geq \cdots \geq 0$ is a nonincreasing sequence of positive terms. Cauchy proved that the original series and the "condensed" series $u_0 + 2u_1 + 4u_3 + 8u_7 + \cdots + 2^k u_{2^k - 1} + \cdots$ converge or diverge together. In this case, selected multiples of a subcollection of terms tell us all we need to know about the behavior of the original infinite series. It seems too good to be true.

We conclude this section with a lesser known convergence test from Cauchy's arsenal, one that demonstrates his endless fascination with this topic [19].

Theorem: If $\sum_{k=1}^{\infty} u_k$ is a series of positive terms for which $\lim\limits_{k \to \infty} \dfrac{\ln(u_k)}{\ln(1/k)} = h > 1$, then the series converges.

Proof: As with the root test, Cauchy sought a "buffer" between 1 and h and so chose a real number a with $1 < a < h$. This guaranteed the existence of a positive integer m so that $\dfrac{\ln(u_k)}{\ln(1/k)} > a$ for all $k \geq m$. From there, he observed that

$$a < \frac{\ln(u_k)}{\ln(1/k)} = \frac{-\ln(u_k)}{\ln k} \quad \text{and so} \quad a \ln(k) < \ln\left(\frac{1}{u_k}\right).$$

Exponentiating both sides of this inequality, he deduced that $k^a < \dfrac{1}{u_k}$

and so $u_k < \dfrac{1}{k^a}$ for all $k \geq m$. But $\displaystyle\sum_{k=m}^{\infty} \frac{1}{k^a}$ (which is now called a p-series)

converges because $a > 1$, and so the original series $\displaystyle\sum_{k=1}^{\infty} u_k$ converges by the comparison test.   Q.E.D.

As an example, consider $\displaystyle\sum_{k=1}^{\infty} \frac{\ln(k)}{k^p}$, where $p > 1$. Cauchy's test requires us

to evaluate $\displaystyle\lim_{k\to\infty} \frac{\ln[\ln(k)/k^p]}{\ln(1/k)}$, which suggests in turn that we first simplify the quotient:

$$\frac{\ln[\ln(k)/k^p]}{\ln(1/k)} = \frac{\ln[\ln(k)] - p\ln(k)}{-\ln(k)} = -\frac{\ln[\ln(k)]}{\ln(k)} + p.$$

By l'Hospital's rule, $\displaystyle\lim_{k\to\infty}\left(-\frac{\ln[\ln(k)]}{\ln(k)} + p\right) = p > 1$, establishing the conver-

gence of $\displaystyle\sum_{k=1}^{\infty} \frac{\ln(k)}{k^p}$ by Cauchy's test. It is a very nice result.

Before leaving Augustin-Louis Cauchy, we offer an apology and a pre-view. We apologize for a chapter that reads like a précis of an introductory analysis text. Indeed, there is no stronger testimonial to Cauchy's influ-ence than that his "greatest hits" are now the heart and soul of the subject. Building upon the idea of limit, he developed elementary real analysis in a way that remains the model to this day. As Bell properly observed, Cauchy stands at center stage, and it is for this reason that the present chapter is one of the book's longest. It could hardly be otherwise.

This brings us to the preview. None of these accolades should sug-gest that, after Cauchy, the quest was finished. On at least three fronts there was still work to be done, work that will occupy us in chapters to come.

First, his definitions could be made more general and his proofs more rigorous. A satisfactory definition of the integral, for instance, need not be limited to continuous functions, and the nagging issue of uniformity had to be identified and resolved. These tasks would fall largely to the German mathematicians Georg Friedrich Bernhard Riemann and Karl Weierstrass, who in a sense supplied the last word on mathematical precision.

Second, Cauchy's more theoretical approach to continuity, differentiability, and integrability motivated those who followed to sort out the connections among these concepts. Such connections would intrigue mathematicians throughout the nineteenth century, and their resulting theorems—and counterexamples—would hold plenty of surprises.

Finally, the need to understand the completeness property raised questions about the very nature of the real numbers. The answers to these questions, combined with the arrival of set theory, would change the face of analysis, although no mathematician active in 1840 could know that a revolution lay just over the horizon.

But *any* mathematician active in 1840 would have known about Cauchy. On this front, we shall give the last word to math historian Carl Boyer. In his classic study of the history of calculus, Boyer wrote, "Through [his] works, Cauchy did more than anyone else to impress upon the subject the character which it bears at the present time" [20].

In a very real sense, all who followed are his disciples.

CHAPTER 7

Riemann

Georg Friedrich Bernhard Riemann

By this point of our story, the "function" had assumed a central importance in analysis. At first it may have seemed like a straightforward, even innocuous notion, but as the collection of functions grew ever more sophisticated—and ever more strange—mathematicians realized they had a conceptual tiger by the tail.

To sketch this evolution, we return briefly to the origins. As we have seen, seventeenth century scholars like Newton and Leibniz believed that the raw material of their new subject was the curve, a concept rooted in the geometric/intuitive approach that later analysts would abandon.

It was largely because of Euler that attention shifted from curves to functions. This significant change in viewpoint, dating from the publication of his *Introductio in analysin infinitorum*, positioned real analysis as the study of functions and their behavior.

Euler addressed this matter early in the *Introductio*. He first distinguished between a *constant* quantity (one that "always keeps the same value") and a *variable* quantity ("one which is not determined or is universal, which can take on any value") and then adopted the following definition: "A function of a variable quantity is an analytic expression composed in any way whatsoever of the variable quantity and numbers or constant quantities" [1]. As examples he offered expressions like $a + 3z, az + b\sqrt{a^2 - z^2}$, and c^z.

These ideas were a huge improvement upon the "curve" and represented a triumph of algebra over geometry. However, his definition identified functions with analytic expressions—which is to say, functions with *formulas*. Such an identification painted mathematicians into some bizarre corners. For instance, the function $f(x) = \begin{cases} x & \text{if } x \geq 0, \\ -x & \text{if } x < 0, \end{cases}$ as shown in figure 7.1 was considered "discontinuous" not because its graph jumped around but because its *formula* did. Of course, it is perfectly continuous by the modern (i.e., Cauchy's) definition. Worse, as Cauchy observed, we could express the same function by a single formula $g(x) = \sqrt{x^2}$.

There seemed to be ample reason to adopt a more liberal, and liberating, view of what a function could be. Euler himself took a step in this direction a few years after providing the definition above. In his 1755 text on differential calculus, he wrote

> Those quantities that depend on others . . . , namely, those that undergo a change when others change, are called functions of

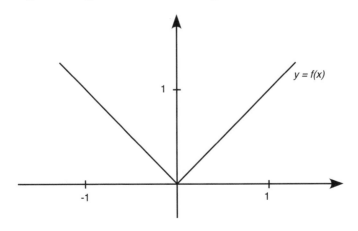

Figure 7.1

these quantities. This definition applies rather widely and includes all ways in which one quantity can be determined by others. [2]

It is important to note that this time he made no explicit reference to analytic expressions, although in his examples of functions Euler retreated to familiar formulas like $y = x^2$.

As the eighteenth century became the nineteenth, functions were revisited in the study of real-world problems about vibrating strings and dissipating heat. This story has been told repeatedly (see, for instance, [3] and [4]), so we note here only that a key figure in the evolving discussion was Joseph Fourier. He came to believe that *any* function defined between $-a$ and a (be it the position of a string, or the distribution of heat in a rod, or something entirely "arbitrary") could be expressed as what we now call a Fourier series:

$$f(x) = \frac{1}{2} a_0 + \sum_{k=1}^{\infty} \left(a_k \cos \frac{n \pi x}{a} + b_k \sin \frac{n \pi x}{a} \right),$$

where the coefficients a_k and b_k are given by

$$a_k = \frac{1}{a} \int_{-a}^{a} f(x) \cos \frac{n \pi x}{a} \, dx \quad \text{and} \quad b_k = \frac{1}{a} \int_{-a}^{a} f(x) \sin \frac{n \pi x}{a} \, dx. \quad (1)$$

To insure that his readers were under no illusions about the level of generality, Fourier explained that his results applied to "a function completely arbitrary, that is to say, a succession of given values, subject or not to a common law," and he went on to describe the values of $y = f(x)$ as succeeding one another "in any manner whatever, and each of them is given as if it were a single quantity" [5].

This statement extended the "late Euler" position that functions could take values at will across different points of their domain. On the other hand, it was by no means clear that the formulas in (1) always hold. The coefficients a_k and b_k are integrals, but how do we know that integrals of general functions even make sense? At least implicitly, Fourier had raised the question of the *existence* of a definite integral, or, in modern terminology, of whether a function is or is not integrable.

As it turned out, Fourier had badly overstated his case, for not every function can be expressed as a Fourier series nor integrated as required by (1). Further, in practice he restricted himself, as had Euler before him, to examples that were fairly routine and well behaved. If the concept of a truly "arbitrary" function were to catch on, someone would have to exhibit one.

DIRICHLET'S FUNCTION

That somebody was Peter Gustav Lejeune-Dirichlet (1805–1859), a gifted mathematician who had studied with Gauss in Germany and with Fourier in France. Over his career, Dirichlet contributed to branches of mathematics ranging from number theory to analysis to that wonderful hybrid of the two called, appropriately enough, analytic number theory.

Here we consider only a portion of Dirichlet's 1829 paper "*Sur la convergence des séries trigonométriques qui servent a représenter une fonction arbitrarie entre des limites données*" (On the Convergence of Trigonometric Series that Represent an Arbitrary Function between Given Limits) [6]. In it, he returned to the representability of functions by a Fourier series like (1) and the implicit existence of those integrals determining the coefficients.

We recall that Cauchy defined his integral for functions continuous on an interval $[\alpha, \beta]$. Using what we now call "improper integrals," Cauchy extended his idea to functions with finitely many points of discontinuity in $[\alpha, \beta]$. For instance, if f is continuous except at a single point r within $[\alpha, \beta]$, as shown in Figure 7.2, Cauchy defined the integral as

$$\int_\alpha^\beta f(x)\,dx = \int_\alpha^r f(x)\,dx + \int_r^\beta f(x)\,dx \equiv \lim_{t \to r^-} \int_\alpha^t f(x)\,dx + \lim_{t \to r^+} \int_t^\beta f(x)\,dx,$$

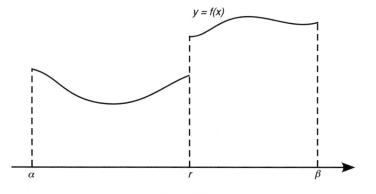

$y = f(x)$

α r β

Figure 7.2

provided all limits exist. If f has discontinuities at $r_1 < r_2 < r_3 < \cdots < r_n$, we define the integral analogously as

$$\int_\alpha^\beta f(x)\,dx \equiv \int_\alpha^{r_1} f(x)\,dx + \int_{r_1}^{r_2} f(x)\,dx + \int_{r_2}^{r_3} f(x)\,dx + \cdots + \int_{r_n}^\beta f(x)\,dx.$$

However, if a function had *infinitely* many discontinuities in the interval $[\alpha, \beta]$, Cauchy's integral was of no use. Dirichlet suggested that a new, more inclusive theory of integration might be crafted to handle such functions, a theory connected to "the fundamental principles of infinitestimal analysis." He never developed ideas in this direction nor did he show how to integrate highly discontinuous functions. He did, however, furnish an example to show that such things exist.

"One supposes," he wrote, "that $\phi(x)$ equals a determined constant c when the variable x takes a rational value and equals another constant d when the variable is irrational" [7]. This is what we now call Dirichlet's function, written concisely as

$$\phi(x) = \begin{cases} c & \text{if } x \text{ is rational,} \\ d & \text{if } x \text{ is irrational.} \end{cases} \tag{2}$$

By the Fourier definition, ϕ was certainly a function: to each x there corresponded one y, even if the correspondence arose from no (obvious) analytic formula. But the function is impossible to graph because of the thorough intermixing of rationals and irrationals on the number line: between any two rationals there is an irrational and vice versa. The graph of ϕ would thus jump back and forth between c and d infinitely often as we move through any interval, no matter how narrow. Such a thing cannot be drawn nor, perhaps, imagined.

Worse, ϕ has no point of continuity. This follows because of the same intermixing of rationals and irrationals. Recall that Cauchy had defined continuity of ϕ at a point x by $\lim_{i \to 0}[\phi(x + i) - \phi(x)] = 0$. As i moves toward 0, it passes through an infinitude of rational and irrational points. As a consequence, $\phi(x + i)$ jumps wildly back and forth, so that the limit in question not only fails to be zero but fails even to *exist*. Because this is the case for any x, the function has no point of continuity.

The significance of this example was twofold. First, it demonstrated that Fourier's idea of an arbitrary function had teeth to it. Before Dirichlet, even those who advocated a more general concept of function had not, in the words of math historian Thomas Hawkins, "taken the implications of this idea seriously" [8]. Dirichlet, by contrast, showed that the world of functions was more vast than anyone had thought. Second, his example suggested an inadequacy in Cauchy's approach to the integral. Perhaps integration could be recast so as not to restrict mathematicians to integrating continuous functions or those with only finitely many discontinuity points.

It was Dirichlet's brilliant student, the abundantly named Georg Friedrich Bernhard Riemann (1826–1866), who took up this challenge. Riemann sought to define the integral without prior assumptions about how continuous a function must be. Divorcing integrability from continuity was a bold and provocative idea.

THE RIEMANN INTEGRAL

In his 1854 *Habilitationsschrift*, a high-level dissertation required of professors at German universities, Riemann stated the issue simply: "What is one to understand by $\int_a^b f(x)dx$?" [9]. Assuming f to be bounded on $[a, b]$, he proceeded with his answer.

First, he took any sequence of values $a < x_1 < x_2 < \cdots < x_{n-1} < b$ within the interval $[a, b]$. Such a subdivision is now called a *partition*. He denoted the lengths of the resulting subintervals by $\delta_1 = x_1 - a$, $\delta_2 = x_2 - x_1$, $\delta_3 = x_3 - x_2$, and so on up to $\delta_n = b - x_{n-1}$. Riemann next let $\varepsilon_1, \varepsilon_2, \ldots, \varepsilon_n$ be a sequence of values between 0 and 1; thus, for each ε_k, the number $x_{k-1} + \varepsilon_k\delta_k$ lies between $x_{k-1} + 0 \cdot \delta_k = x_{k-1}$ and $x_{k-1} + 1 \cdot \delta_k = x_{k-1} + (x_k - x_{k-1}) = x_k$. In other words, $x_{k-1} + \varepsilon_k\delta_k$ falls within the subinterval $[x_{k-1}, x_k]$. He then introduced

$$S = \delta_1 f(a + \varepsilon_1\delta_1) + \delta_2 f(x_1 + \varepsilon_2\delta_2) + \delta_3 f(x_2 + \varepsilon_3\delta_3)$$
$$+ \cdots + \delta_n f(x_{n-1} + \varepsilon_n\delta_n).$$

The reader will recognize this as what we now (appropriately) call a Riemann sum. As illustrated in figure 7.3, it is the total of the areas of rectangles standing upon the various subintervals, where the kth rectangle has base δ_k and height $f(x_{k-1} + \varepsilon_k\delta_k)$.

Riemann was now ready with his critical definition:

> If this sum has the property that, however the δ_k and ε_k are chosen, it becomes infinitely close to a fixed value A as the δ_k become infinitely small, then we call this fixed value $\int_a^b f(x)dx$. If the sum does not have this property, then $\int_a^b f(x)dx$ has no meaning [10].

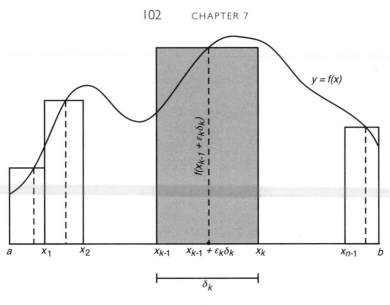

Figure 7.3

This is the first appearance of the Riemann integral, now featured promi-
nently in any course in calculus and, most likely, in any introduction to
real analysis. It is evident that this definition assumed *nothing* about continuity. For Riemann, unlike for Cauchy, continuity was a nonissue.

Returning to the function f and the partition $a < x_1 < x_2 < \cdots < x_{n-1} < b$,
Riemann introduced D_1 as the "greatest oscillation" of the function between
a and x_1. In his words, D_1 was "the difference between the greatest and
least values [of f] in this interval." Similarly, $D_2, D_3, \ldots, D_n$ were the greatest oscillations of f over the subintervals $[x_1, x_2], [x_2, x_3], \ldots, [x_{n-1}, b]$,
and he let D be the difference between the maximum and minimum val-
ues of f over the entire interval $[a, b]$. Clearly $D_k \leq D$, because f cannot
oscillate more over a subinterval than it does across all of $[a, b]$.

A modern mathematician would define these oscillations with more
care. Because f is assumed to be bounded, we know from the all-important
completeness property that the set of real numbers $\{f(x) \mid x \in [x_{k-1}, x_k]\}$
has both a least upper bound and a greatest lower bound. We then let D_k
be the difference of these. In the mid-nineteenth century, however, this
approach would not have been feasible, for the concepts of a least upper
bound and a greatest lower bound—now called, respectively, a *supremum*
and an *infimum*—rested upon vague geometrical intuition if they were
perceived at all.

Be that as it may, Riemann introduced the new sum

$$R = \delta_1 D_1 + \delta_2 D_2 + \delta_3 D_3 + \cdots + \delta_n D_n. \tag{3}$$

R is the shaded area, determined by the difference between the function's largest and smallest values over each subinterval, shown in figure 7.4.

He next let $d > 0$ be a positive number and looked at all partitions of $[a, b]$ for which max $\{\delta_1, \delta_2, \delta_3, \ldots, \delta_n\} \le d$. In words, he was considering those partitions for which even the widest subinterval is of length d or less. Reverting to modern terminology, we define the *norm* of a partition to be the width of the partition's biggest subinterval, so Riemann was here looking at all partitions with norm less than or equal to d. He then introduced $\Delta = \Delta(d)$ to be the "greatest value" of all sums R in (3) arising from partitions with norm less than or equal to d. (Today we would define $\Delta(d)$ as a supremum.)

It was clear to Riemann that the integral $\int_a^b f(x)dx$ existed if and only if $\lim_{d \to 0} \Delta(d) = 0$. Geometrically, this means that as we take increasingly fine partitions of $[a, b]$, the largest shaded area in figure 7.4 will decrease to zero.

He then posed the critical question, "In which cases does a function allow integration and in which does it not?" As before, he was ready with an answer—what we now call the Riemann integrability condition—although the notational baggage became even heavier. Because of the importance of these ideas to the history of analysis, we follow along a little further.

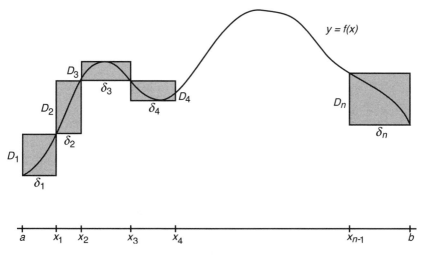

Figure 7.4

First, he let $\sigma > 0$ be a positive number. For a given partition, he looked at those subintervals for which the oscillation of the function was greater than σ. To illustrate, we refer to figure 7.5, where we display the function, its shaded rectangles, and a value of σ at the left. Comparing σ to the heights of the rectangles, we see that on only the two subintervals $[x_1, x_2]$ and $[x_4, x_5]$ does the oscillation exceed σ. We shall call these "Type A" subintervals. The others, where the oscillation is less than or equal to σ, we call "Type B" subintervals. In figure 7.5, the subintervals of Type B are $[a, x_1]$, $[x_2, x_3]$, $[x_3, x_4]$, and $[x_5, b]$.

As a last convention, Riemann let $s = s(\sigma)$ be the *combined* length of the Type A subintervals for a given σ; that is, $s(\sigma) \equiv \sum_{\text{Type A}} \delta_k$. For our example, $s(\sigma) = (x_2 - x_1) + (x_5 - x_4)$. With this notation behind him, Riemann was now ready to prove a necessary and sufficient condition that a bounded function on $[a, b]$ be integrable.

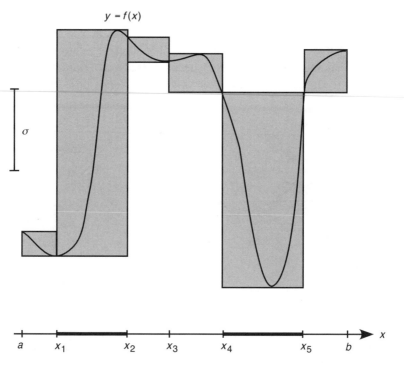

$y = f(x)$

σ

a x_1 x_2 x_3 x_4 x_5 b x

Figure 7.5

Riemann Integrability Condition: $\int_a^b f(x)dx$ exists if and only if, for any $\sigma > 0$, the combined length of the Type A subintervals can be made as small as we wish by letting $d \to 0$.

Admittedly, there is a lot going on here. In words, this says that f is integrable if and only if, for any σ no matter how small, we can find a norm so that, for all partitions of $[a, b]$ having a norm that small or smaller, the total length of the subintervals where the function oscillates more than σ is negligible. We examine Riemann's necessity and sufficiency proofs separately.

Necessity: If $\int_a^b f(x)dx$ exists and we fix a value of $\sigma > 0$, then $\lim\limits_{d \to 0} s(\sigma) = 0$.

Proof: Riemann began with a partition of unspecified norm d and considered $R = \delta_1 D_1 + \delta_2 D_2 + \delta_3 D_3 + \cdots + \delta_n D_n$ from (3). He noted that $R \geq \sum\limits_{\text{Type A}} \delta_k D_k$, because the summation on the right includes the Type A terms and omits the others. But for each Type A subinterval, the oscillation of f exceeds σ; this is, of course, how the Type A subintervals are identified in the first place. So, recalling the definition of $s(\sigma)$, we have

$$R \geq \sum_{\text{Type A}} \delta_k D_k \geq \sum_{\text{Type A}} \delta_k \sigma = \sigma \cdot \sum_{\text{Type A}} \delta_k = \sigma \cdot s(\sigma).$$

On the other hand, $R = \delta_1 D_1 + \delta_2 D_2 + \delta_3 D_3 + \cdots + \delta_n D_n \leq \Delta(d)$ because $\Delta(d)$ is the greatest such value for all partitions having norm d or less.
    Riemann combined this pair of inequalities to get $\sigma \cdot s(\sigma) \leq R \leq \Delta(d)$. Ignoring the middle term and dividing by σ, he concluded that

$$0 \leq s(\sigma) \leq \frac{\Delta(d)}{\sigma}. \tag{4}$$

Recall that, in proving necessity, he had assumed that f is integrable, and this in turn meant that $\Delta(d) \to 0$ as $d \to 0$. Because σ was a fixed number, $\dfrac{\Delta(d)}{\sigma} \to 0$ as well. It follows from (4) that, as d approaches zero, the value of $s(\sigma)$ must likewise go to zero.   Q.E.D.

This was the conclusion Riemann sought: that the total length $s(\sigma)$ of subintervals where the function oscillates more than σ can be made, as he wrote, "arbitrarily small with suitable values of d." That was half the battle. Next in line was the converse.

Sufficiency: If for any $\sigma > 0$, we have $\lim\limits_{d \to 0} s(\sigma) = 0$, then $\int_a^b f(x)dx$ exists.

Proof: This time Riemann began by noting that, for any $\sigma > 0$, we have

$$R = \delta_1 D_1 + \delta_2 D_2 + \delta_3 D_3 + \cdots + \delta_n D_n = \underset{\text{Type A}}{\sum \delta_k D_k} + \underset{\text{Type B}}{\sum \delta_k D_k}. \quad (5)$$

Here he simply broke the summation into two parts, depending on whether the interval was of Type A (where the function oscillates more than σ) or of Type B (where it does not). He then treated these summands separately.

For the first, he recalled that $D_k \leq D$, where D was the oscillation of f over the entire interval $[a, b]$. Thus,

$$\underset{\text{Type A}}{\sum \delta_k D_k} \leq \underset{\text{Type A}}{\sum \delta_k D} = D \cdot \underset{\text{Type A}}{\sum \delta_k} = D \cdot s(\sigma). \quad (6)$$

Meanwhile, for each Type B subinterval we know that $D_k \leq \sigma$, and so

$$\underset{\text{Type B}}{\sum \delta_k D_k} \leq \underset{\text{Type B}}{\sum \delta_k \sigma} = \sigma \cdot \underset{\text{Type B}}{\sum \delta_k} \leq \sigma \cdot \sum_{k=1}^n \delta_k = \sigma(b - a), \quad (7)$$

where we have replaced the sum of the lengths of the Type B subintervals with the larger value $b - a$, the sum of the lengths of *all* the subintervals. Riemann now assembled (5), (6), and (7) to get the inequality

$$R = \underset{\text{Type A}}{\sum \delta_k D_k} + \underset{\text{Type B}}{\sum \delta_k D_k} \leq Ds(\sigma) + \sigma(b - a). \quad (8)$$

Because (8) holds for any positive σ, we can fix a value of σ so that $\sigma(b - a)$ is as small as we wish. For this fixed value of σ, we recall the hypothesis that as $d \to 0$, then $s(\sigma)$ goes to zero as well. We thus can choose d so that $Ds(\sigma)$ is also small. From (8) it follows that the corresponding values of R can be made arbitrarily small, and so the *greatest* of these—what Riemann called $\Delta(d)$—will likewise be arbitrarily small. This meant that

$\lim\limits_{d \to 0} \Delta(d) = 0$, which was Riemann's way of saying that f is integrable on $[a, b]$.   Q.E.D.

This complicated argument has been taken intact from Riemann's 1854 paper. Although notationally intricate, the fundamental idea is simple: in order for a function to have a Riemann integral, its oscillations must be under control. A function that jumps too often and too wildly cannot be integrated. From a geometrical viewpoint, such a function would seem to have no definable area beneath it.

The Riemann integrability condition is a handy device for showing when a bounded function is or is not integrable. Consider again Dirichlet's function in (2). For the sake of specificity, we take $c = 1$ and $d = 0$ and restrict our attention to the unit interval $[0, 1]$. Then we have

$$\phi(x) = \begin{cases} 1 & \text{if} \quad x \text{ is rational,} \\ 0 & \text{if} \quad x \text{ is irrational.} \end{cases}$$

The question is whether, by Riemann's definition, the integral $\int_0^1 \phi(x)dx$ exists.

As we have seen, the integrability condition replaces this question by one involving oscillations of the function. Suppose we let $\sigma = 1/2$ and consider any partition $0 < x_1 < x_2 < \cdots < x_{n-1} < 1$ and any resulting subinterval $[x_k, x_{k+1}]$. Because this subinterval, no matter how narrow, contains infinitely many rationals and infinitely many irrationals, the oscillation of ϕ on $[x_k, x_{k+1}]$ is $1 - 0 = 1 > 1/2 = \sigma$. As a consequence, *every* subinterval of the partition is of Type A, and so $s(1/2) = \sum\limits_{\text{Type A}} \delta_k = 1$, the entire length of $[0, 1]$. In short, $s(1/2) = 1$ for *any* partition of $[0, 1]$.

Riemann's condition required that, for ϕ to be integrable, $s(1/2) = \sum\limits_{\text{Type A}} \delta_k$ can be made as small as we wish by choosing suitably fine partitions $[0,1]$. But as we have seen, the value of $s(1/2)$ is 1 no matter how we tinker with the partition, so we surely cannot make it less than, say, 0.01. Because the integrability condition cannot be met, this function is not integrable. According to Riemann, $\int_0^1 \phi(x)dx$ is nonsense.

Intuitively, Dirichlet's function is so thoroughly discontinuous that it cannot be integrated. This phenomenon raised a fundamental question: just how discontinuous can a function be and still be integrable by Riemann's definition? Although this mystery would not be solved until the twentieth century, Riemann himself described a function that provided a tantalizing piece of evidence.

RIEMANN'S PATHOLOGICAL FUNCTION

As noted, Riemann introduced no prior assumptions about continuity and thereby suggested that some very bizarre functions—those that "are discontinuous infinitely often," as he put it—might be integrated. "As these functions are as yet nowhere considered," he wrote, "it will be good to provide a specific example" [11].

First he let $(x) = x - n$, where n is the integer nearest to x. Thus, $(1.2) = (-1.8) = 0.2$, whereas $(1.7) = (-1.3) = -0.3$. If x fell halfway between two integers, like 4.5 or -0.5, then he set $(x) = 0$. The graph of $y = (x)$ appears in figure 7.6. It is clear that the function has a jump discontinuity of length 1 at each $x = \pm m/2$, where m is an odd whole number.

Riemann next considered $y = (2x)$, which "compressed" figure 7.6 horizontally and resulted in the graph of figure 7.7. Here jumps of length 1 occur at $x = \pm m/4$, where m is an odd whole number.

This compression process continued with $y = (3x)$, $y = (4x)$, and so on, until Riemann assembled these into the function of interest:

$$f(x) = \frac{(x)}{1} + \frac{(2x)}{4} + \frac{(3x)}{9} + \frac{(4x)}{16} + \cdots = \sum_{k=1}^{\infty} \frac{(kx)}{k^2}.$$

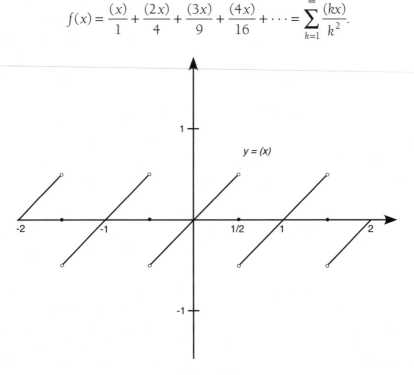

Figure 7.6

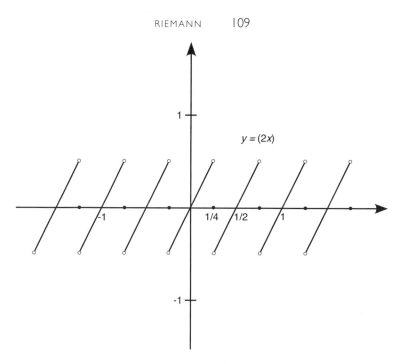

$y = (2x)$

Figure 7.7

To get a sense of f, we have graphed its seventh partial sum, that is,
$$\frac{(x)}{1} + \frac{(2x)}{4} + \frac{(3x)}{9} + \frac{(4x)}{16} + \frac{(5x)}{25} + \frac{(6x)}{36} + \frac{(7x)}{49}, \text{ over the interval } [0, 1]$$
in figure 7.8. Even at this stage, it appears that the discontinuities of f are fast accumulating.

We observe that $|(kx)| \leq \frac{1}{2}$ for all x, and so the infinite series converges everywhere by a comparison test with $\sum_{k=1}^{\infty} \frac{1}{2k^2}$. Riemann asserted, without a complete proof, that f is continuous at those points where each individual function $y = (kx)$ is continuous, and this would include all the irrationals. But he also asserted that, if $x = \frac{m}{2n}$, where m and n are relatively prime integers, then f has a jump at x of length $\frac{1}{n^2}\left(1 + \frac{1}{9} + \frac{1}{25} + \frac{1}{49} + \frac{1}{81} + \cdots\right) = \frac{\pi^2}{8n^2}$. (Here we have summed the series using Euler's result from chapter 4.)

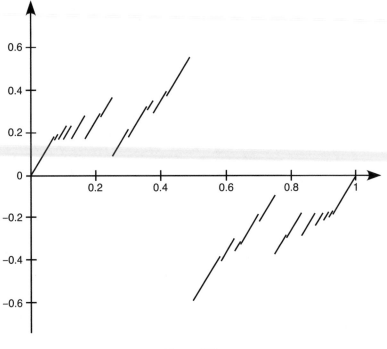

Figure 7.8

Thus, Riemann's function has discontinuities at points like $\frac{55}{14}$ or $\frac{-3}{38}$ or $\frac{81}{1000}$. There are infinitely many such points between any two real numbers, and so his function had infinitely many points of discontinuity within any finite interval. This should meet anyone's criterion for "highly discontinuous."

Nonetheless—and this is the amazing part—$\int_0^1 f(x)dx$ exists. Riemann proved this by means of the integrability condition above. He began with an arbitrary $\sigma > 0$, although to simplify our discussion, we shall specify $\sigma = \frac{1}{20}$. We must identify those points where the oscillation of the function exceeds $\frac{1}{20}$, and these are rationals of the form $x = \frac{m}{2n}$. But the size of the jump at such points is $\frac{\pi^2}{8n^2}$, so we need only consider the inequality $\frac{\pi^2}{8n^2} > \frac{1}{20}$. It follows that $n < \frac{\pi}{2}\sqrt{10} \approx 4.967$, and because n is a whole number, the only options are $n = 1, 2, 3,$ or 4. When we note as well that m and n have no common factors and that $0 \le \dfrac{m}{2n} \le 1$, we conclude that there are only *finitely* many such candidates. In this case, the points in $[0, 1]$ where the function oscillates more than $\dfrac{1}{20}$ are: $\dfrac{1}{8}, \dfrac{1}{6}, \dfrac{1}{4}, \dfrac{1}{3}, \dfrac{3}{8}, \dfrac{1}{2}, \dfrac{5}{8}, \dfrac{2}{3}, \dfrac{3}{4}, \dfrac{5}{6},$ and $\dfrac{7}{8}$.

Because we have only finitely many points to deal with, we can create a partition of $[0, 1]$ that places each of these within a very narrow subinterval, the total length of which can be as small as we wish. For instance, to include the eleven points above in subintervals with *total* length less than $1/100$, we might begin our partition with

$$0 < x_1 = \frac{1}{8} - \frac{1}{10000} = \frac{1249}{10000} < x_2 = \frac{1}{8} + \frac{1}{10000} = \frac{1251}{10000},$$

thereby embedding the discontinuity at $x = \dfrac{1}{8}$ in a subinterval of total length $\delta_1 = \dfrac{1251}{10000} - \dfrac{1249}{10000} = \dfrac{1}{5000}$. If we put equally narrow intervals about each of the Type A points for $\sigma = \dfrac{1}{20}$, then $s\left(\dfrac{1}{20}\right) = 11 \times \left(\dfrac{1}{5000}\right) < \dfrac{1}{100}$.

The critical issue here is the *finite* number of points where the oscillation exceeds a given σ. Riemann summarized the situation as follows: "In all intervals which do not contain these jumps, the oscillations are less than σ and . . . the total length of the intervals that contain these jumps can, at our pleasure, be made small" [12].

Riemann had constructed a function with infinitely many discontinuities in any interval yet that met his integrability condition. It was a peculiar creation, one that is now known as Riemann's pathological function, where the adjective carries the connotation of being, in some sense, "sick."

Of course, Riemann had not answered the question, "How discontinuous can an integrable function be?" But he had shown that integrable functions could be stunningly discontinuous. To those critics who sneered that an example as weird as Riemann's was of no practical use, he offered a persuasive rejoinder: "This topic stands in the closest association with the principles of infinitesimal analysis and can serve to bring to these principles greater clarity and precision. In this respect, the topic has an immediate interest" [13]. Riemann's pathological function had precisely this effect, even if it did provide a blow to the mathematical intuition. As we shall see, more intuition-busters were in store for analysts of the nineteenth century.

THE RIEMANN REARRANGEMENT THEOREM

To be sure, Riemann is best known for his theory of the integral, but we end this chapter in a different corner of analysis, with a Riemannian result that may be less important than whimsical, but one that never ceases to amaze the first-time student.

We begin by recalling the Leibniz series from chapter 2, namely, $1 - \dfrac{1}{3} + \dfrac{1}{5} - \dfrac{1}{7} + \dfrac{1}{9} - \cdots$. Suppose we rearrange the terms of this series in the following manner: take the first two positive terms followed by the first negative; take the next two positive terms followed by the second negative; and so on. After grouping this rearrangement into threesomes, we have

$$\left(1 + \frac{1}{5} - \frac{1}{3}\right) + \left(\frac{1}{9} + \frac{1}{13} - \frac{1}{7}\right) + \left(\frac{1}{17} + \frac{1}{21} - \frac{1}{11}\right) + \left(\frac{1}{25} + \frac{1}{29} - \frac{1}{15}\right) + \cdots.$$

$$(9)$$

A moment's thought reveals that the expressions in parentheses look like

$$\frac{1}{8k - 7} + \frac{1}{8k - 3} - \frac{1}{4k - 1} \quad \text{for } k = 1, 2, 3, 4, \ldots,$$

and these can be combined into $\dfrac{24k - 11}{(8k - 7)(8k - 3)(4k - 1)}.$

Because $k \geq 1$, both the numerator and denominator of this last fraction must be positive, and so the value of each threesome in (9) will be positive as well. We thus can say the following about the rearranged series:

$$\left(1 + \frac{1}{5} - \frac{1}{3}\right) + \left(\frac{1}{9} + \frac{1}{13} - \frac{1}{7}\right) + \left(\frac{1}{17} + \frac{1}{21} - \frac{1}{11}\right) + \left(\frac{1}{25} + \frac{1}{29} - \frac{1}{15}\right) + \cdots$$

$$\geq \left(1 + \frac{1}{5} - \frac{1}{3}\right) + 0 + 0 + 0 + 0 + \cdots = \frac{13}{15} = 0.8666\ldots.$$

On the other hand, Leibniz had proved that the original series $1 - \frac{1}{3} + \frac{1}{5} - \frac{1}{7} + \frac{1}{9} - \cdots = \frac{\pi}{4} \approx 0.7854$. We are left with an inescapable conclusion: the rearranged series, whose sum has been shown to exceed 0.8666, cannot converge to the same number as the original. By altering not the terms of the series but their *position*, we have changed the sum. This seems mighty odd.

Actually, it gets worse, for Riemann showed how the Leibniz series can be rearranged to converge to any number at all!

His reasoning is expedited by the introduction of some terminology and a few well-known theorems. As we saw, it was Cauchy who said what it means for an infinite series $\sum_{k=1}^{\infty} u_k$ to converge. A general series may, of course, include both positive and negative terms, and this suggests that we disregard the signs and look at $\sum_{k=1}^{\infty} |u_k|$ instead. If this latter series converges, we say that $\sum_{k=1}^{\infty} u_k$ converges *absolutely*. If $\sum_{k=1}^{\infty} u_k$ converges but $\sum_{k=1}^{\infty} |u_k|$ does not, the original series is said to converge *conditionally*.

As an example, we return to the original series of Leibniz. It sums to $\frac{\pi}{4}$, but the related series of absolute values diverges because

$$1 + \frac{1}{3} + \frac{1}{5} + \frac{1}{7} + \frac{1}{9} + \cdots \geq \frac{1}{2} + \frac{1}{4} + \frac{1}{6} + \frac{1}{8} + \frac{1}{10} + \cdots$$

$$= \frac{1}{2}\left[1 + \frac{1}{2} + \frac{1}{3} + \frac{1}{4} + \frac{1}{5} + \cdots\right],$$

where we recognize the divergent harmonic series in the brackets. This means that Leibniz's series is conditionally convergent.

It is customary when dealing with series of mixed signs to consider the positives and the negatives separately. Following Riemann's notation, we write a series as $(a_1 + a_2 + a_3 + a_4 + \cdots) + (-b_1 - b_2 - b_3 - b_4 - \cdots)$, where all the a_k and b_k are nonnegative. Riemann knew that if the original series converged absolutely, then both of the series $\sum_{k=1}^{\infty} a_k$ and $\sum_{k=1}^{\infty} b_k$ converge; if the original series diverged, then one of $\sum_{k=1}^{\infty} a_k$ and $\sum_{k=1}^{\infty} b_k$

diverges to infinity; and if the original converged conditionally, then both $\sum_{k=1}^{\infty} a_k$ and $\sum_{k=1}^{\infty} b_k$ diverge to infinity.

It was Dirichlet who showed that any rearrangement of an absolutely convergent series must converge to the same sum as the original [14]. For absolutely convergent series, repositioning the terms has no impact whatever.

But for conditionally convergent series, we reach a dramatically different conclusion: if a series converges conditionally, it can be rearranged to converge to whatever number we wish. With some alliterative excess, we might call this Riemann's remarkable rearrangement result. Here is the idea of his proof.

Letting C be a fixed number—our "target," so to speak—Riemann began thus: "One alternately takes sufficiently many positive terms of the series that their sum exceeds C and then sufficiently many negative terms that the (combined) sum is less than C" [15]. To see what he was getting at, we stipulate that our target C is positive. Starting with the positive terms, we find the smallest m so that $a_1 + a_2 + a_3 + \cdots + a_m > C$. There surely is such an index because $\sum_{k=1}^{\infty} a_k$ diverges to infinity. One next considers the negative terms and chooses the smallest n so that $a_1 + a_2 + a_3 + \cdots + a_m - b_1 - b_2 - \cdots - b_n < C$. Again, we know such an index exists because the divergent series $\sum_{k=1}^{\infty} b_k$ must eventually exceed $(a_1 + a_2 + a_3 + \cdots + a_m) - C$. But $a_1 + a_2 + a_3 + \cdots + a_m - b_1 - b_2 - \cdots - b_n$ is a rearrangement of terms of the original series whose sum can be no further from C than b_n. The process is then repeated, adding some a_k and subtracting some b_k so that the difference between C and this sum of these rearranged terms is less than some b_p. Because the original series converges, we know its general term goes to zero, so $\lim_{r \to \infty} b_r = 0$ as well. The series rearranged by his alternating scheme will converge to C as claimed. It is quite wonderful.

To illustrate, suppose we sought a rearrangement of Leibniz's series that would converge to, say, 1.10. We would begin with sufficiently many positive terms to exceed this: $1 + \dfrac{1}{5} = 1.2 > 1.10$. Then we would subtract a negative term to bring us below 1.10:

$$\left(1 + \frac{1}{5}\right) - \frac{1}{3} = 0.8666 \cdots < 1.10.$$

Then we add back some positive terms until we again surpass 1.10, then bounce back with some negatives, and so on. With this recipe, the rearranged Leibniz series that converges to 1.10 will begin as follows:

$$\left(1+\frac{1}{5}\right)-\frac{1}{3}+\left(\frac{1}{9}+\frac{1}{13}+\frac{1}{17}\right)-\frac{1}{7}+\left(\frac{1}{21}+\frac{1}{25}+\frac{1}{29}+\frac{1}{33}\right)-\frac{1}{11}+\cdots.$$

Once seen, Riemann's argument seems self-evident. Nonetheless, his rearrangement theorem demonstrates in dramatic fashion that summing infinite series is a tricky business. By simply rearranging the terms we can drastically alter the answer. As has been observed previously, the study of infinite processes, which is to say analysis, can carry us into deep waters.

With that, we leave Georg Friedrich Bernhard Riemann, although no journey through nineteenth century analysis can leave him for long. More than anyone, he established the integral as a primary player in the calculus enterprise. And his ideas would serve as the point of departure for Henri Lebesgue, who, as we shall see in the book's final chapter, picked up where Riemann left off to develop his own revolutionary theory of integration.

CHAPTER 8

Liouville

Joseph Liouville

Generality lies at the heart of modern analysis, a trend already evident in the limit theorems of Cauchy or the integrals of Riemann. More than their predecessors, these mathematicians defined key concepts inclusively and drew conclusions valid not for one or two cases but for enormous families. It was a most significant development.

Yet the century witnessed another, seemingly opposite, phenomenon: the growing importance of the explicit example and the specific counterexample. These deserve our attention alongside the general theorems of the preceding pages. In this chapter, we examine Joseph Liouville's discovery of the first transcendental number in 1851; in the next, we consider Karl Weierstrass's astonishingly pathological function from 1872. Each of these was a major achievement of its time, and each reminds us that analysis would be incomplete without the clarification provided by individual examples.

To study transcendentals, we need some background on where the problem originated, how it was refined over the decades, and why its resolution was such a grand achievement. We start, as did calculus itself, in the seventeenth century.

THE ALGEBRAIC AND THE TRANSCENDENTAL

It appears to have been Leibniz who first used the term "transcendental" in a mathematical classification scheme. Writing about his newly invented differential calculus, Leibniz noted its applicability to fractions, roots, and similar algebraic quantities, but then added, "It is clear that our method also covers transcendental curves—those that cannot be reduced by algebraic computation or have no particular degree—and thus holds in a most general way"[1]. Here Leibniz wanted to separate those entities that were algebraic, and thus reasonably straightforward, from those that were intrinsically more sophisticated.

The distinction was refined by Euler in the eighteenth century. In his *Introductio*, he listed the so-called algebraic operations as "addition, subtraction, multiplication, division, raising to a power, and extraction of roots," as well as "the solution of equations." Any other operations were transcendental, such as those involving "exponentials, logarithms, and others which integral calculus supplies in abundance" [2]. He even went so far as to mention transcendental *quantities* and gave as an example "logarithms of numbers that are not powers of the base," although he provided no airtight definition nor rigorous proof [3].

Our mathematical forebears had the right idea, even if they failed to express it precisely. To them it was evident that certain mathematical objects, be they curves, functions, or numbers, were accessible via the fundamental operations of algebra, whereas others were sufficiently complicated to transcend algebra altogether and thereby earn the name "transcendental."

After contributions from such late eighteenth century mathematicians as Legendre, an unambiguous definition appeared. A real number was said to be *algebraic* if it solved some polynomial equation with integer coefficients. That is, x_0 is an algebraic number if there exists a polynomial $P(x) = ax^n + bx^{n-1} + cx^{n-2} + \cdots + gx + h$, where $a, b, c, \ldots, g$, and h are integers and such that $P(x_0) = 0$. For instance, $\sqrt{2}$ is algebraic because it is a solution of $x^2 - 2 = 0$, a quadratic equation with integer coefficients. Less obviously, the number $\sqrt{2} + \sqrt[3]{5}$ is algebraic for it solves $x^6 - 6x^4 - 10x^3 + 12x^2 - 60x + 17 = 0$.

From a geometric perspective, an algebraic number is the x-intercept of the graph of $y = P(x)$, where P is a polynomial with integer coefficients (see figure 8.1). If we imagine graphing on the same axes all linear, all quadratic, all cubic—generally all polynomials whose coefficients are integers—then the infinite collection of their x-intercepts will be the algebraic numbers.

An obvious question arises: Is there anything else? To allow for this possibility, we say a real number is *transcendental* if it is not algebraic. Any real number must, by sheer logic, fall into one category or the other.

But are there any transcendentals? A piece of terminology, after all, does not guarantee existence. A mammalogist might just as well define a dolphin to be algebraic if it lives in water and to be transcendental if it does not. Here, the concept of a transcendental dolphin is unambiguous, but no such thing exists.

Mathematicians had to face a similar possibility. Could transcendental numbers be a well-defined figment of the imagination? Might all those (algebraic) x-intercepts cover the line completely? If not, where should one look for a number that is not the intercept of *any* polynomial equation with integer coefficients?

As a first step toward an answer, we note that a transcendental number must be irrational. For, if $x_0 = a/b$ is rational, then x_0 obviously satisfies the first-degree equation $bx - a = 0$, whose coefficients b and $-a$ are integers. Indeed, the rationals are precisely those algebraic numbers satisfying linear equations with integer coefficients.

Of course, not every algebraic number is rational, as is clear from the algebraic irrationals $\sqrt{2}$ and $\sqrt{2} + \sqrt[3]{5}$. Algebraic numbers thus represent a generalization of the rationals in that we now drop the requirement that they solve polynomials of the *first* degree (although we retain the restriction that coefficients be integers).

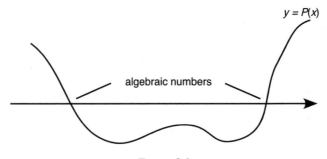

$y = P(x)$

algebraic numbers

Figure 8.1

Transcendentals, if they exist, must lurk among the irrationals. From the time of the Greeks, roots like $\sqrt{2}$ were known to be irrational, and by the end of the eighteenth century, the irrationality of the constants e and π had been established, respectively, by Euler in 1737 and Johann Lambert (1728–1777) in 1768 [4]. But proving irrationality is a far easier task than proving transcendence.

As we noted, Euler conjectured that the number $\log_2 3$ is transcendental, and Legendre believed that π was as well [5]. However, beliefs of mathematicians, no matter how fervently held, prove nothing. Deep into the nineteenth century, the existence of even a single transcendental number had yet to be demonstrated. It remained possible that these might occupy the same empty niche as those transcendental dolphins.

An example was provided at long last by the French mathematician Joseph Liouville (1809–1882). Modern students may remember his name from Sturm–Liouville theory in differential equations or from Liouville's theorem ("an entire, bounded function is constant") in complex analysis. He contributed significantly to such applied areas as electricity and thermodynamics and, in an entirely different arena, was elected to the Assembly of France during the tumultuous year of 1848. On top of all of this, for thirty-nine years he edited one of the most influential journals in the history of mathematics, originally titled *Journal de mathématiques pures et appliquées* but often referred to simply as the *Journal de Liouville*. In this way, he was responsible for transmitting mathematical ideas to colleagues around Europe and the world [6].

Within real analysis, Liouville is remembered for two significant discoveries. First was his proof that certain elementary functions cannot have elementary antiderivatives. Anyone who has taken calculus will remember applying clever schemes to find indefinite integrals. Although these matters are no longer addressed with quite as much zeal as in the past, calculus courses still cover techniques like integration by parts and integration by partial fractions that allow us to compute such antiderivatives as $\int x^2 e^{-x} dx = -x^2 e^{-x} - 2xe^{-x} - 2e^{-x} + C$ or the considerably less self-evident

$$\int \sqrt{\tan x}\, dx = \frac{1}{\sqrt{8}} \ln \left| \frac{\tan x - \sqrt{2\tan x} + 1}{\tan x + \sqrt{2\tan x} + 1} \right|$$
$$+ \frac{1}{\sqrt{2}} \arctan \left(\frac{\sqrt{2\tan x}}{1 - \tan x} \right) + C.$$

Note that both the integrands and their antiderivatives are composed of functions from the standard Eulerian repertoire: algebraic, trigonometric, logarithmic, and their inverses. These are "elementary" integrals with "elementary" antiderivatives.

Alas, even the most diligent integrator will be stymied in his or her quest for $\int \sqrt{\sin x}\, dx$ as a finite combination of simple functions. It was Liouville who proved in an 1835 paper why a closed-form answer for certain integrals is impossible. For instance, he wrote that, "One easily convinces oneself by our method that the integral $\int \dfrac{e^x}{x}\, dx$, which has greatly occupied geometers, is impossible in finite form" [7]. The hope that easy functions must have easy antiderivatives was destroyed forever.

In this chapter our object is Liouville's other famous contribution: a proof that transcendental numbers exist. His original argument came in 1844, although he refined and simplified the result in a classic 1851 paper (published in his own journal, of course) from which we take the proof that follows [8]. Before providing his example of a hitherto unseen transcendental, Liouville first had to prove an important inequality about irrational algebraic numbers and their rational neighbors.

LIOUVILLE'S INEQUALITY

As noted, a real number is algebraic if it is the solution to some polynomial equation with integer coefficients. Any number that solves one such equation, however, solves infinitely many. For instance, $\sqrt{2}$ is the solution of the quadratic equation $x^2 - 2 = 0$, as well as the cubic equation $x^3 + x^2 - 2x - 2 = (x^2 - 2)(x + 1) = 0$, the quartic equation $x^4 + 4x^3 + x^2 - 8x - 6 = (x^2 - 2)(x + 1)(x + 3) = 0$, and so on. Our first stipulation, then, is that we use a polynomial of minimal degree. So, for the algebraic number $\sqrt{2}$, we would employ the quadratic above and not its higher degree cousins.

Suppose that x_0 is an *irrational* algebraic number. Following Liouville's notation, we denote its minimal-degree polynomial by

$$P(x) = ax^n + bx^{n-1} + cx^{n-2} + \cdots + gx + h, \qquad (1)$$

where $a, b, c, \ldots, g$, and h are integers and $n \geq 2$ (as noted above, if $n = 1$, the algebraic number is rational). Because $P(x_0) = 0$, the factor theorem allows us to write

$$P(x) = (x - x_0)\, Q(x), \qquad (2)$$

where Q is a polynomial of degree $n - 1$. Liouville wished to establish a bound upon the size of $|Q(x)|$, at least for values of x in the vicinity of x_0. We give his proof and then follow it with a simpler alternative.

Liouville's Inequality: If x_0 is an irrational algebraic number with minimum-degree polynomial $P(x) = ax^n + bx^{n-1} + cx^{n-2} + \cdots + gx + h$ having integer coefficients and degree $n \geq 2$, then there exists a positive real number A so that, if p/q is a rational number in $[x_0 - 1, x_0 + 1]$, then

$$\left| \frac{p}{q} - x_0 \right| \geq \frac{1}{Aq^n}.$$

Proof: The argument has its share of fine points, but we begin with the real polynomial Q introduced in (2). This is continuous and thus bounded on any closed, finite interval, so there exists an $A > 0$ with

$$|Q(x)| \leq A \quad \text{for all } x \text{ in } [x_0 - 1, x_0 + 1]. \tag{3}$$

Now consider any rational number p/q within one unit of x_0, where we insist that the rational be in lowest terms and that its denominator be positive (i.e., that $q \geq 1$). We see by (3) that $|Q(p/q)| \leq A$. We claim as well that $P(p/q) \neq 0$, for otherwise we could factor $P(x) = \left(x - \dfrac{p}{q} \right) R(x)$, and it can be shown that R will be an $(n-1)$st-degree polynomial having integer coefficients. Then $0 = P(x_0) = \left(x_0 - \dfrac{p}{q} \right) R(x_0)$ and yet $\left(x_0 - \dfrac{p}{q} \right) \neq 0$ (because the rational p/q differs from the irrational x_0), and we would conclude that $R(x_0) = 0$. This, however, makes x_0 a root of R, a polynomial with integer coefficients having lower degree than P, in violation of the assumed minimality condition. It follows that p/q is not a root of $P(x) = 0$.

Liouville returned to the minimal-degree polynomial in (1) and defined $f(p,q) \equiv q^n P(p/q)$. Note that

$$\begin{aligned}
f(p,q) &= q^n P(p/q) \\
&= q^n [a(p/q)^n + b(p/q)^{n-1} + c(p/q)^{n-2} + \cdots + g(p/q) + h] \\
&= ap^n + bp^{n-1}q + cp^{n-2}q^2 + \cdots + gpq^{n-1} + hq^n. \tag{4}
\end{aligned}$$

From (4), he made a pair of simple but telling observations.

First, $f(p, q)$ is an integer, for its components $a, b, c, \ldots, g, h$, along with p and q, are all integers. Second, $f(p, q)$ cannot be zero, for, if $0 = f(p, q) = q^n P(p/q)$, then either $q = 0$ or $P(p/q) = 0$. The former is impossible because q is a denominator, and the latter is impossible by our discussion above. Thus, Liouville knew that $f(p, q)$ was a *nonzero* integer, from which he deduced that

$$|q^n \, P(p/q)| = |f(p, q)| \geq 1. \tag{5}$$

The rest of the proof followed quickly. From (3) and (5) and the fact that $P(x) = (x - x_0) \, Q(x)$, he concluded that

$$1 \leq |q^n P(p/q)| = q^n |p/q - x_0||Q(p/q)| \leq q^n |p/q - x_0|A.$$

Hence $|p/q - x_0| \geq 1/Aq^n$, and the demonstration was complete. Q.E.D.

The role played by inequalities in Liouville's proof is striking. Modern analysis is sometimes called the "science of inequalities," a characterization that is appropriate here and would become ever more so as the century progressed.

We promised an alternate proof of Liouville's result. This time, our argument features Cauchy's mean value theorem in a starring role [9].

Liouville's Inequality Revisited: If x_0 is an irrational algebraic number with minimum-degree polynomial $P(x) = ax^n + bx^{n-1} + cx^{n-2} + \cdots + gx + h$ having integer coefficients and degree $n \geq 2$, then there exists an $A > 0$ such that, if p/q is a rational number in $[x_0 - 1, x_0 + 1]$, then,

$$\left| \frac{p}{q} - x_0 \right| \geq \frac{1}{Aq^n}.$$

Proof: Differentiating P, we find $P'(x) = nax^{n-1} + (n-1)bx^{n-2} + (n-2)cx^{n-3} + \cdots + g$. This $(n-1)$st-degree polynomial is bounded on $[x_0 - 1, x_0 + 1]$, so there is an $A > 0$ for which $|P'(x)| \leq A$ for all $x \in [x_0 - 1, x_0 + 1]$. Letting p/q be a rational number within one unit of x_0 and applying the mean value theorem to P, we know there exists a point c between x_0 and p/q for which

$$\frac{P(p/q) - P(x_0)}{p/q - x_0} = P'(c). \tag{6}$$

Given that $P(x_0) = 0$ and c belongs to $[x_0 - 1, x_0 + 1]$, we see from (6) that

$$|P(p/q)| = |p/q - x_0| \cdot |P'(c)| \le A|p/q - x_0|.$$

Consequently, $|q^n P(p/q)| \le Aq^n|p/q - x_0|$. But, as noted above, $q^n P(p/q)$ is a nonzero integer, and so $1 \le Aq^n|p/q - x_0|$. The result follows. Q.E.D.

At this point, an example might be of interest. We consider the algebraic irrational $x_0 = \sqrt{2}$. Here the minimal-degree polynomial is $P(x) = x^2 - 2$, the derivative of which is $P'(x) = 2x$. It is clear that, on the interval $[\sqrt{2} - 1, \sqrt{2} + 1]$, P' is bounded by $A = 2\sqrt{2} + 2$. Liouville's inequality shows that, if p/q is any rational in this closed interval, then $\left| \dfrac{p}{q} - \sqrt{2} \right| \ge$

$\dfrac{1}{(2\sqrt{2} + 2)q^2}$.

The numerically inclined may wish to verify this for, say, $q = 5$. In this case, the inequality becomes $\left| \dfrac{p}{5} - \sqrt{2} \right| \ge \dfrac{1}{(50\sqrt{2} + 50)} \approx 0.00828$. We then check all the "fifths" within one unit of $\sqrt{2}$. Fortunately, there are only ten such fractions, and all abide by Liouville's inequality:

$p/5$	$|p/5 - \sqrt{2}|$
$3/5 = 0.60$	0.8142
$4/5 = 0.80$	0.6142
$5/5 = 1.00$	0.4142
$6/5 = 1.20$	0.2142
$7/5 = 1.40$	0.0142
$8/5 = 1.60$	0.1858
$9/5 = 1.80$	0.3858
$10/5 = 2.00$	0.5858
$11/5 = 2.20$	0.7858
$12/5 = 2.40$	0.9858

The example suggests something more: we can in general remove the restriction that p/q lies close to x_0. That is, we specify A^* to be the greater of 1 and A, where A is determined as above. If p/q is a rational within one unit of x_0, then

$$\left| \frac{p}{q} - x_0 \right| \ge \frac{1}{Aq^n} \ge \frac{1}{A^* q^n} \quad \text{because } A^* \ge A.$$

On the other hand, if p/q is a rational *more* than one unit away from x_0, then

$$\left| \frac{p}{q} - x_0 \right| \geq 1 \geq \frac{1}{A^*} \geq \frac{1}{A^* q^n} \quad \text{because } A^* \geq 1 \text{ and } q \geq 1 \text{ as well.}$$

The upshot of this last observation is that there exists an $A^* > 0$ for which

$$\left| \frac{p}{q} - x_0 \right| \geq \frac{1}{A^* q^n} \quad \text{regardless of the proximity of } p/q \text{ to } x_0.$$

Informally, Liouville's inequality shows that rational numbers are poor approximators of irrational algebraics, for there must be a gap of at least $\dfrac{1}{A^* q^n}$ between x_0 and any rational p/q. It is not easy to imagine how Liouville noticed this. That he did so, and offered a clever proof, is a tribute to his mathematical ability. Yet all may have been forgotten had he not taken the next step: he used his result to find the world's first transcendental.

LIOUVILLE'S TRANSCENDENTAL NUMBER

We first offer a word about the logical strategy. Liouville sought an irrational number that was *inconsistent* with the conclusion of the inequality above. This irrational would thus violate the inequality's assumptions, which means it would not be algebraic. If Liouville could pull this off, he would have corralled a specific transcendental. Remarkably enough, he did just that [10].

Theorem: The real number $x_0 \equiv \displaystyle\sum_{k=1}^{\infty} \frac{1}{10^{k!}} = \frac{1}{10} + \frac{1}{10^2} + \frac{1}{10^6} + \frac{1}{10^{24}} +$ $\dfrac{1}{10^{120}} + \cdots$ is transcendental.

Proof: There are three issues to address, and we treat them one at a time.

First, we claim that the series defining x_0 is convergent, and this follows easily from the comparison test. That is, $k! \geq k$ guarantees that

$$\frac{1}{10^{k!}} \leq \frac{1}{10^k}, \text{ and so } \sum_{k=1}^{\infty} \frac{1}{10^{k!}} \text{ converges because } \sum_{k=1}^{\infty} \frac{1}{10^k} = \frac{1/10}{1 - 1/10} = \frac{1}{9}.$$

In short, x_0 is a real number.

Second, we assert that x_0 is irrational. This is clear from its decimal expansion, $0.1100010000000 \ldots$, where nonzero entries occupy the first place, the second, the sixth, the twenty-fourth, the one-hundred twentieth, and so on, with ever-longer strings of 0s separating the increasingly lonely 1s. Obviously no finite block of this decimal expansion repeats, so x_0 is irrational.

The final step is the hardest: to show that Liouville's number is transcendental. To do this, we assume instead that x_0 is an algebraic irrational with minimal polynomial of degree $n \geq 2$. By Liouville's inequality, there must exist an $A^* > 0$ such that, for any rational p/q, we have $\left| \dfrac{p}{q} - x_0 \right| \geq \dfrac{1}{A^* q^n}$ and, as a consequence,

$$0 < \frac{1}{A^*} \leq q^n \left| \frac{p}{q} - x_0 \right|. \tag{7}$$

We now choose an arbitrary whole number $m > n$ and look at the partial sum $\displaystyle\sum_{k=1}^{m} \frac{1}{10^{k!}} = \frac{1}{10} + \frac{1}{10^2} + \frac{1}{10^6} + \cdots + \frac{1}{10^{m!}}$. If we combine these fractions, their common denominator would be $10^{m!}$, so we could write the sum as $\displaystyle\sum_{k=1}^{m} \frac{1}{10^{k!}} = \frac{p_m}{10^{m!}}$, where p_m is a whole number. Thus, of course, $\dfrac{p_m}{10^{m!}}$ is a rational.

Comparing this to x_0, we see that

$$\left| \frac{p_m}{10^{m!}} - x_0 \right| = \sum_{k=m+1}^{\infty} \frac{1}{10^{k!}} = \frac{1}{10^{(m+1)!}} + \frac{1}{10^{(m+2)!}} + \frac{1}{10^{(m+3)!}} + \cdots .$$

An induction argument establishes that $(m+r)! \geq (m+1)! + (r-1)$ for any whole number $r \geq 1$, and so $\dfrac{1}{10^{(m+r)!}} \leq \dfrac{1}{10^{(m+1)!+r-1}} = \dfrac{1}{10^{(m+1)!}} \left[\dfrac{1}{10^{r-1}} \right]$. As a consequence,

$$\left| \frac{p_m}{10^{m!}} - x_0 \right| = \frac{1}{10^{(m+1)!}} + \frac{1}{10^{(m+2)!}} + \frac{1}{10^{(m+3)!}} + \cdots$$

$$\leq \frac{1}{10^{(m+1)!}} + \frac{1}{10^{(m+1)!} \times 10} + \frac{1}{10^{(m+1)!} \times (10^2)}$$

$$+ \frac{1}{10^{(m+1)!} \times (10^3)} + \cdots$$

$$= \frac{1}{10^{(m+1)!}} \left[1 + \frac{1}{10} + \frac{1}{100} + \frac{1}{1000} + \cdots \right]$$

$$= \frac{1}{10^{(m+1)!}} \left[\frac{10}{9} \right] < \frac{2}{10^{(m+1)!}}. \tag{8}$$

A contradiction is now at hand because

$$0 < \frac{1}{A^*} \le (10^{m!})^n \left| \frac{p_m}{10^{m!}} - x_0 \right| \qquad \text{by (7)}$$

$$< (10^{m!})^n \cdot \frac{2}{10^{(m+1)!}} \qquad \text{by (8)}$$

$$= \frac{2}{10^{(m+1)!-n(m!)}} = \frac{2}{10^{m!(m+1-n)}} < \frac{2}{10^{m!}},$$

where the last step follows because $m > n$ implies that $m + 1 - n > 1$.

This long string of inequalities shows that, for the value of A^* introduced above, we have $\frac{1}{A^*} < \frac{2}{10^{m!}}$ for all $m > n$, or simply that $2A^* > 10^{m!}$ for all $m > n$. Such an inequality is absurd, for $2A^*$ is a fixed number, whereas $10^{m!}$ explodes to infinity as m gets large. Liouville had (at last) reached a contradiction.

By this time, the reader may need a gentle reminder of what was contradicted. It was the assumption that the irrational x_0 is algebraic. There remains but one alternative: x_0 must be transcendental. And the existence of such a number is what Joseph Liouville had set out to prove.   Q.E.D.

In his 1851 paper, Liouville observed that, although many had speculated on the existence of transcendentals, "I do not believe a proof has ever been given" to this end [11]. Now, one had.

Strangely enough, Liouville regarded this achievement as something less than a total success, for his original hope had been to show that the number e was transcendental [12]. It is one thing to *create* a number, as Liouville did, and then prove its transcendence. It is quite another to do this for a number like e that was "already there." With his typical flair, Eric Temple Bell observed that it is

a much more difficult problem to prove that a *particular* suspect, like e or π, is or is not transcendental than it is to invent a whole infinite class of transcendentals: ... the suspected number is entire master of the situation, and it is the mathematician in this case, not the suspect, who takes orders. [13]

We might say that Liouville demonstrated the transcendence of a number no one had previously cared about but was unable to do the same for the ubiquitous constant e, about which mathematicians cared passionately.

Still, it would be absurd to label him a failure when he found something his predecessors had been seeking in vain for a hundred years.

That original objective would soon be realized by one of his followers. In 1873, Charles Hermite (1822–1901) showed that e was indeed a transcendental number. Nine years later Ferdinand Lindemann (1852–1939) proved the same about π. As is well known, the latter established the impossibility of squaring the circle with compass and straightedge, a problem with origins in classical Greece that had gone unresolved not just for decades or centuries but for *millennia* [14]. The results of Hermite and Lindemann were impressive pieces of reasoning that built upon Liouville's pioneering research.

To this day, determining whether a given number is transcendental ranks among the most difficult challenges in mathematics. Much work has been done on this front and many important theorems have been proved, but there remain vast holes in our understanding. Among the great achievements, we should mention the 1934 proof of A. O. Gelfond (1906–1968), which demonstrated the transcendence of an entire family of numbers at once. He proved that if a is an algebraic number other than 0 or 1 and if b is an *irrational* algebraic, then a^b must be transcendental. This deep result guarantees, for instance, that $2^{\sqrt{2}}$ or $(\sqrt{2} + \sqrt[3]{5})^{\sqrt{7}}$ are transcendental. Among other candidates now known to be transcendental are e^{π}, $\ln(2)$, and $\sin(1)$.

However, as of this writing, the nature of such "simple" numbers as π^e, e^e, and π^{π} is yet to be established. Worse, although mathematicians believe in their bones that both $\pi + e$ and $\pi \times e$ are transcendental, no one has actually proved this [15]. We repeat: demonstrating transcendence is very, very hard.

Returning to the subject at hand, we see how far mathematicians had come by the mid-nineteenth century. Liouville's technical abilities in manipulating inequalities as well as his broader vision of how to attack so difficult a problem are impressive indeed. Analysis was coming of age.

Yet this proof will serve as a dramatic counterpoint to our main theorem from chapter 11. There, we shall see how Georg Cantor found a remarkable shortcut to reach Liouville's conclusion with a fraction of the work. In doing so, he changed the direction of mathematical analysis. The Liouville–Cantor interplay will serve as a powerful reminder of the continuing vitality of mathematics.

For now, Cantor must wait a bit. Our next object is the ultimate in nineteenth century rigor: the mathematics of Karl Weierstrass and the greatest analytic counterexample of all.

CHAPTER 9

Weierstrass

Karl Weierstrass

As we have seen, mathematicians of the nineteenth century imparted to the calculus a new level of rigor. By our standards, however, these achievements were not beyond criticism. Reading mathematics from that period is a bit like listening to Chopin performed on a piano with a few keys out of tune: one can readily appreciate the genius of the music, yet now and then something does not quite ring true. The modern era would not arrive until the last vestige of imprecision disappeared and analytic arguments became, for all practical purposes, incontrovertible. The mathematician most responsible for this final transformation is Karl Weierstrass (1815–1897).

He followed a nontraditional route to prominence. His student years had been those of an underachiever, featuring more beer and swordplay than is normally recommended. At age 30 Weierstrass found himself on the faculty of a German *gymnasium* (i.e., high school) far removed from the intellectual centers of Europe. By day, he instructed his pupils on the arts of arithmetic and calligraphy, and only after classes were finished and the lessons corrected could young Weierstrass turn to his research [1].

In 1854 this unknown teacher from an unknown town published a memoir on Abelian integrals that astonished the mathematicians who read it. It was evident that the author, whoever he was, possessed an extraordinary talent. Within two years, Weierstrass had secured a position at the University of Berlin and found himself on one of the world's great mathematics faculties. His was a true Cinderella story.

Weierstrass's contributions to analysis were as profound as his pedagogical skills were legendary. With a reputation that spread through Germany and beyond, he attracted young mathematicians who wished to learn from the master. A school of disciples formed at his feet. This was almost literally true, for severe vertigo required Weierstrass to lecture from an easy chair while a designated student wrote his words upon the board (an arrangement subsequent professors have envied but seldom replicated).

If his teaching style was unusual, so was his attitude toward publication. Although his classes were filled with new and important ideas, he often let others disseminate such information in their own writings. Thus one finds his results attributed somewhat loosely to the School of Weierstrass. Modern academics, operating in "publish or perish" mode, find it difficult to fathom such a nonpossessive view of scholarship. But Weierstrass acted as though *creating* significant mathematics was his job, and he would risk the perishing.

Whether through his own publications or those of his lieutenants, the Weierstrassian school imparted to analysis an unparalleled logical precision. He repaired subtle misconceptions, proved important theorems, and constructed a counterexample that left mathematicians shaking their heads. In this chapter, we shall see why Karl Weierstrass came to be known, in the parlance of the times, as the "father of modern analysis" [2].

BACK TO THE BASICS

We recall that Cauchy built his calculus upon limits, which he defined in these words:

> When the values successively attributed to a variable approach indefinitely to a fixed value, in a manner so as to end by differing from it by as little as one wishes, this last is called the limit of all the others.

To us, aspects of this statement, for instance, the motion implied in the term "approach," seem less than satisfactory. Is something actually moving? If so, must we consider concepts of time and space before talking of limits? And what does it mean for the process to "end"? The whole business needed one last revision.

Contrast Cauchy's words with the polished definition from the Weierstrassians:

$$\lim_{x \to a} f(x) = L \text{ if and only if, for every } \varepsilon > 0, \text{ there exists a } \delta > 0$$

$$\text{so that, if } 0 < |x - a| < \delta, \text{ then } |f(x) - L| < \varepsilon. \tag{1}$$

Here nothing is in motion, and time is irrelevant. This is a static rather than dynamic definition and an arithmetic rather than a geometric one. At its core, it is nothing but a statement about inequalities. And it can be used as the foundation for unambiguous proofs of limit theorems, for example, that the limit of a sum is the sum of the limits. Such theorems could now be demonstrated with all the rigor of a proposition from Euclid.

Some may argue that precision comes at a cost, for Weierstrass's austere definition lacks the charm of intuition and the immediacy of geometry. To be sure, a statement like (1) takes some getting used to. But geometrical intuition was becoming suspect, and this purely analytic definition was in no way entangled with space or time.

Besides reformulating key concepts, Weierstrass grasped their meanings as his predecessors had not. An example is uniform continuity, a property that Cauchy missed entirely. We recall that Cauchy defined continuity on a point-by-point basis, saying that f is continuous at a if $\lim_{x \to a} f(x) = f(a)$. In Weierstrassian language, this means that to every $\varepsilon > 0$, there corresponds a $\delta > 0$ so that, if $0 < |x - a| < \delta$, then $|f(x) - f(a)| < \varepsilon$. Thus, for a fixed "target" ε and a given a, we can find the necessary δ. But here δ depends on *both* ε and a. Were we to keep the same ε but consider a different value of a, the choice of δ would, in general, have to be adjusted.

It was Eduard Heine (1821–1881) who first drew this distinction in print, although he suggested that "the general idea" was conveyed to him by his mentor, Weierstrass [3]. Heine defined a function f to be *uniformly continuous* on its domain if, for every $\varepsilon > 0$, there exists a $\delta > 0$ so that, if x and y are any two points in the domain within δ units of one another, then $|f(x) - f(y)| < \varepsilon$. This means, in essence, that "one δ fits all," so that points within this uniform distance will have functional values within ε of one another.

It is clear that a uniformly continuous function will be continuous at each individual point. The converse, however, is false, and the standard counterexample is the function $f(x) = 1/x$ defined on the open interval $(0, 1)$, as shown in figure 9.1. This is certainly continuous at each point of $(0, 1)$, but it fails Heine's criterion for uniformity. To see why, we let $\varepsilon = 1$ and claim that there can be no $\delta > 0$ with the property that, when x and y are chosen from $(0, 1)$ with $|x - y| < \delta$, then $|f(x) - f(y)| = \left|\dfrac{1}{x} - \dfrac{1}{y}\right| < 1$.

For, given any proposed δ, we can choose an integer $N > \max\{1/\delta, 1\}$ and let $x = 1/(N + 2)$ and $y = 1/N$. In this case, both x and y belong to $(0,1)$ and

$$|x - y| = \left|\frac{1}{N+2} - \frac{1}{N}\right| = \frac{2}{N(N+2)} < \frac{N+2}{N(N+2)} = \frac{1}{N} < \delta.$$

But $\left|\dfrac{1}{x} - \dfrac{1}{y}\right| = \left|\dfrac{1}{1/(N+2)} - \dfrac{1}{1/N}\right| = 2 \nless 1 = \varepsilon$. The requirement for uniform continuity is not met.

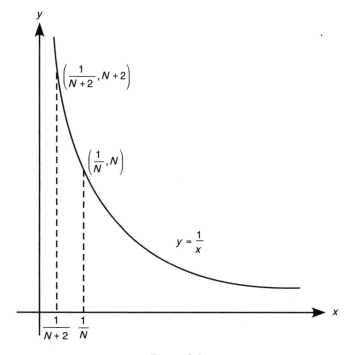

Figure 9.1

A look back to chapter 6 reminds us that Cauchy had talked about continuous functions but actually used *uniform* continuity in some of his proofs. Fortunately, a logical catastrophe was averted in 1872 when Heine proved that a function continuous on a closed, bounded interval $[a, b]$ must in fact be uniformly continuous. That is, the distinction between continuity and uniform continuity disappears if we restrict our attention to $[a, b]$. (Note that the example above is defined on an open interval.) So, when Cauchy's misconception occurred for functions on closed, bounded intervals, his proofs were "salvageable" thanks to Heine's result.

Weierstrass recognized an even more crucial dichotomy: that between pointwise and uniform convergence. These ideas warrant a brief digression.

Suppose we have a sequence of *functions*, $f_1, f_2, f_3, \ldots, f_k, \ldots$, all with the same domain. If we fix a point x in this domain and substitute it into each function, we generate a sequence of *numbers*: $f_1(x), f_2(x), f_3(x), \ldots, f_k(x), \ldots$. Assume that, for each individual x, this numerical sequence converges. We then create a new function f defined at each point x by $f(x) \equiv \lim_{k \to \infty} f_k(x)$. We call f the "pointwise limit" of the f_k.

For instance, consider the following sequence of functions on $[0, \pi]$:
$$f_1(x) = \sin x, \quad f_2(x) = (\sin x)^2, \quad f_3(x) = (\sin x)^3, \ldots, \quad f_k(x) = (\sin x)^k, \ldots,$$
the first three of which are graphed in figure 9.2.

We see that, for all $k \geq 1$, $f_k\left(\dfrac{\pi}{2}\right) = \left(\sin \dfrac{\pi}{2}\right)^k = 1$, and so $\lim_{k \to \infty} f_k\left(\dfrac{\pi}{2}\right) = \lim_{k \to \infty} 1 = 1$. On the other hand, if x is in $[0, \pi]$ but $x \neq \dfrac{\pi}{2}$, then $\sin x = r$, where $0 \leq r < 1$, and so $\lim_{k \to \infty} f_k(x) = \lim_{k \to \infty} (r^k) = 0$. Hence the pointwise limit is

$$f(x) = \lim_{k \to \infty} f_k(x) = \begin{cases} 0 & \text{if } 0 \leq x < \pi/2, \\ 1 & \text{if } x = \pi/2, \\ 0 & \text{if } \pi/2 < x \leq \pi, \end{cases}$$

whose graph is shown in figure 9.3.

This example raises one of the great questions of analysis: if each of the f_k has a certain property and f is their pointwise limit, must f itself have this property? In mathematical parlance, we ask whether a characteristic is *inherited* by pointwise limits. If each f_k is continuous, must f be continuous? If each is integrable, must f be integrable?

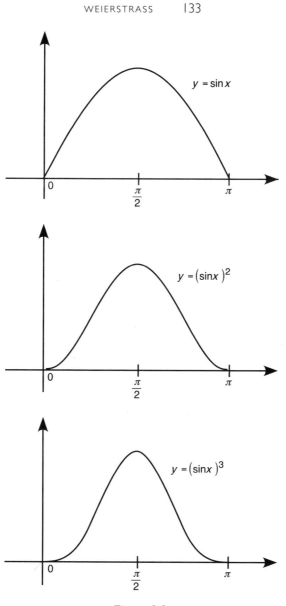

Figure 9.2

The intuitive answer might be, "Sure, why not?" Alas, the world is not so simple. For instance, continuity is not inherited by pointwise limits, a source of confusion for Cauchy and other mathematicians of the past [4]. We need only look at the example above to see that the functions $f_k(x) = (\sin x)^k$ are

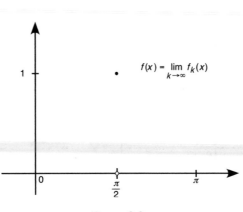

Figure 9.3

continuous everywhere, but their pointwise limit f in figure 9.3 is not continuous at $x = \pi/2$. This same example shows that differentiability is not inherited either.

What about integrals? Already in this book we have seen occasions where mathematicians assumed that

$$\lim_{k \to \infty} \int_a^b f_k(x)\,dx = \int_a^b \left[\lim_{k \to \infty} f_k(x) \right] dx.$$

This asserts that we may safely interchange two important calculus operations: integrate and then take the limit or take the limit and then integrate.

To see that this too is in error, we define a sequence of functions f_k on $[0, 1]$ by

$$f_k(x) = \begin{cases} 0 & \text{if} \quad 0 \le x < \dfrac{1}{2k}, \\[2mm] (16k^2)x - 8k & \text{if} \quad \dfrac{1}{2k} \le x < \dfrac{3}{4k}, \\[2mm] (-16k^2)x + 16k & \text{if} \quad \dfrac{3}{4k} \le x < \dfrac{1}{k}, \\[2mm] 0 & \text{if} \quad \dfrac{1}{k} \le x \le 1. \end{cases}$$

Although this expression may look daunting, the graphs of f_1, f_2, and f_3 in figure 9.4 reveal that the functions are fairly tame. Each is continuous, with "spikes" of increasing height but decreasing width situated ever closer to the origin.

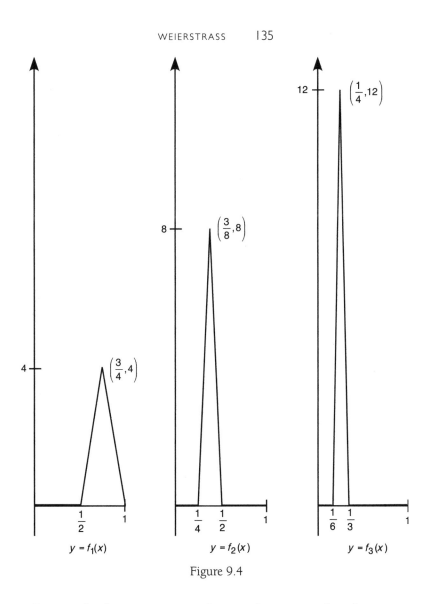

Figure 9.4

Because the f_k are continuous, they can be integrated, and it is easy to evaluate their integrals as triangular areas (see figure 9.5):

$$\int_0^1 f_k(x)\,dx = \text{Area of triangle} = \frac{1}{2}b \times h = \frac{1}{2}\left(\frac{1}{2k}\right) \times (4k) = 1.$$

So, as the bases of these triangular regions get smaller, their heights grow in such a way that the triangular areas remain constant. Clearly, then,

$$\lim_{k\to\infty}\int_0^1 f_k(x)\,dx = \lim_{k\to\infty} 1 = 1. \tag{2}$$

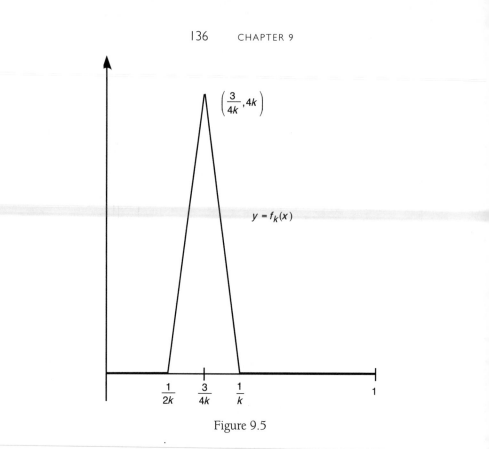

Figure 9.5

On the other hand, we assert that the pointwise limit of the f_k is zero everywhere on [0, 1]. Certainly $f(0) = 0$, because $f_k(0) = 0$ for each k. And if $0 < x \leq 1$, we choose a whole number N so that $\dfrac{1}{N} < x$ and observe that for all subsequent functions, that is, for all f_k with $k \geq N$, the "spike" has moved to the left of x, making $f_k(x) = 0$. Thus $f(x) \equiv \lim_{k \to \infty} f_k(x) = 0$ as well. As a consequence, we see that

$$\int_0^1 \left[\lim_{k \to \infty} f_k(x) \right] dx = \int_0^1 f(x)\, dx = \int_0^1 0 \cdot dx = 0.$$

Comparing this to (2) reveals the disheartening fact that the limit of the integrals need not be the integral of the limits. Symbolically, we have a case where $\lim_{k \to \infty} \int_0^1 f_k(x)\, dx \neq \int_0^1 \left[\lim_{k \to \infty} f_k(x) \right] dx$. Again, pointwise limits do not behave "nicely"—an analytic circumstance much to be regretted.

By 1841 Weierstrass understood this state of affairs and proposed a way around it [5]. Characteristically, he did not publish his ideas until

1894—more than half a century later—but his students had spread the word long before. The idea was to introduce a stronger form of convergence, called *uniform convergence*, under which key properties transfer from individual functions to their limit.

Following his lead, we define a sequence of functions f_k to converge uniformly to a function f on a common domain if for every $\varepsilon > 0$, there is a whole number N so that, if $k \geq N$ and if x is any point in the domain, then $|f_k(x) - f(x)| < \varepsilon$. In a manner reminiscent of uniform continuity, this says that "one N fits all x" in the domain of the functions f_k.

This mode of convergence can be illustrated geometrically. Given $\varepsilon > 0$, we draw a band of width ε surrounding the graph of $y = f(x)$, as shown in figure 9.6. By uniform convergence, we must reach a subscript N so that f_N and all subsequent functions in the sequence lie *entirely* within this band. As the name suggests, such functions approximate f uniformly across the interval $[a, b]$.

It is easy to see that if a sequence of functions converges uniformly to f, then it converges pointwise to f, but not conversely. For example, the "spike" functions described above converge pointwise but not uniformly to the zero function on $[0, 1]$. Uniform convergence is a stronger, more restrictive phenomenon than mere pointwise convergence.

We have undertaken this digression for a few reasons. First, we shall need the notion of uniform convergence in the chapter's main result. Second, echoes of these ideas appear throughout the remainder of the book. Finally, such considerations illustrate why Weierstrass is so important in the history of calculus. In the words of Victor Katz,

> Not only did Weierstrass make absolutely clear how certain quantities in his definition(s) depended on other quantities, but he also completed the transformation away from the use of terms such as "infinitely small." Henceforth, all definitions involving such ideas were given arithmetically [6].

FOUR GREAT THEOREMS

Besides revisiting definitions, Weierstrass was a master at employing them to prove theorems of importance. Here we shall mention (but not prove) four of his results involving uniform convergence.

The first two address a topic mentioned above: under uniform convergence, important analytic properties transfer from the individual f_k to the limit function f.

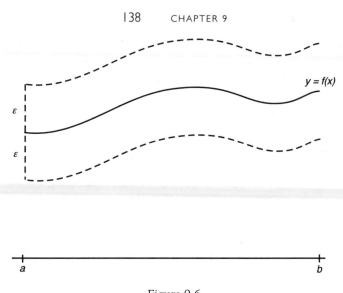

Figure 9.6

Theorem 1: If $\{f_k\}$ is a sequence of continuous functions converging uniformly to f on $[a, b]$, then f itself is continuous.

Theorem 2: If $\{f_k\}$ is a sequence of bounded, Riemann-integrable functions converging uniformly to f on $[a, b]$, then f is Riemann-integrable on $[a, b]$ and

$$\lim_{k \to \infty} \left[\int_a^b f_k(x)\,dx \right] = \int_a^b \left[\lim_{k \to \infty} f_k(x) \right] dx = \int_a^b f(x)\,dx.$$

By theorem 2, the interchange of limits and integrals is permissible for uniformly converging sequences of functions.

The third result is now called the Weierstrass approximation theorem. It provides a fortuitous connection between continuous functions and polynomials.

Theorem 3 (Weierstrass approximation theorem): If f is a continuous function defined on a closed, bounded interval $[a, b]$, then there exists a sequence of polynomials P_k converging uniformly to f on $[a, b]$.

What is so fascinating about this theorem is that continuous functions can be quite ill behaved (this, in fact, is the point of Weierstrass's counterexample, which we examine in a moment). Polynomials, by contrast, are as tame as can be. That the latter uniformly approximate the former seems a wonderful piece of good fortune.

These three theorems, then, make the case for uniform convergence. They allow for the transfer of continuity and integrability from individual functions to their limit and provide a vehicle for approximating continuous functions by polynomials. But is there an easy way to establish uniform convergence in the first place?

One route is to apply the so-called Weierstrass M-test, the last of our preliminary results. As before, we begin with a sequence of functions $\{f_k\}$ defined on a common domain, but the M-test introduces a new twist: we *add* these to create partial sums $S_n(x) = \sum_{k=1}^{n} f_k(x) = f_1(x) + f_2(x) + \cdots + f_n(x)$. If the sequence of partial sums $\{S_n\}$ converges uniformly to a function f, we say the infinite series of functions $\sum_{k=1}^{\infty} f_k(x)$ converges uniformly to f. With this background, we now state the following result.

Theorem 4 (Weierstrass M-test): If a sequence $\{f_k\}$ of functions defined on a common domain has the property that, for each k, there exists a positive number M_k so that $|f_k(x)| \le M_k$ for all x in the domain and if the infinite series $\sum_{k=1}^{\infty} M_k$ converges, then the series of functions $\sum_{k=1}^{\infty} f_k(x)$ converges uniformly.

This amounts to a comparison test between functions and numbers, where convergence of the series of *numbers* implies uniform convergence of the series of *functions*. For example, consider the function defined on [0, 1] by

$$f(x) = \sum_{k=1}^{\infty} \frac{x^k}{(k+1)^3} = \frac{x}{2^3} + \frac{x^2}{3^3} + \frac{x^4}{4^3} + \cdots .$$

Here we have $|f_k(x)| = \left| \frac{x^k}{(k+1)^3} \right| \le \frac{1}{(k+1)^3} \le \frac{1}{k^2}$ for all x in [0,1], and we know that $\sum_{k=1}^{\infty} \frac{1}{k^2} = \frac{\pi^2}{6}$ by Euler's result from chapter 4. Uniform convergence follows immediately from the M-test. Moreover, if we apply theorems

1 and 2 to the partial sums S_n, we know that f is itself continuous because each of the partial sums is and that

$$\int_0^1 f(x)\,dx = \int_0^1 \left[\lim_{n\to\infty} S_n(x)\,dx \right] = \lim_{n\to\infty} \left[\int_0^1 S_n(x)\,dx \right]$$

$$= \lim_{n\to\infty} \left[\int_0^1 \left(\sum_{k=1}^n \frac{x^k}{(k+1)^3} \right) dx \right]$$

$$= \lim_{n\to\infty} \left[\sum_{k=1}^n \left(\int_0^1 \frac{x^k}{(k+1)^3}\,dx \right) \right] = \lim_{n\to\infty} \left[\sum_{k=1}^n \frac{1}{(k+1)^4} \right]$$

$$= \sum_{k=1}^\infty \frac{1}{(k+1)^4} = \left[\sum_{k=1}^\infty \frac{1}{k^4} \right] - 1 = \frac{\pi^4}{90} - 1,$$

again with a little help from Euler. Here we have included all the intervening steps as a reminder of how complicated matters become when we interchange infinite processes. The Weierstrass M-test has allowed us to conclude that f is continuous and to evaluate its integral exactly—a pretty significant accomplishment.

At last the preliminaries are behind us, and the stage is set for a mathematical bombshell.

WEIERSTRASS'S PATHOLOGICAL FUNCTION

Mathematicians long knew that a differentiable ("smooth") function must be continuous ("unbroken"), but not conversely. A V-shaped function like $y = |x|$, for instance, is everywhere continuous but is not differentiable at $x = 0$, where its graph abruptly changes direction to produce a corner.

It was believed, however, that continuous functions must be smooth "most of the time." The renowned André-Marie Ampère (1775–1836) had presented a proof that continuous functions are differentiable in general, and calculus textbooks throughout the first half of the nineteenth century endorsed this position [7].

It certainly has appeal. Anyone can imagine a continuous "sawtooth" graph rising smoothly to a corner, then descending to the next corner, then rising to the next, and so on. As we compress the "teeth," we get ever more points of nondifferentiability. Nonetheless, it seems that there must remain intervals where the graph rises or falls smoothly to get from one corner to the next. In this way, the geometry suggests that any continuous function must have plenty of points of differentiability.

It was thus a shock when Weierstrass constructed his function continuous at every point but differentiable at none, a bizarre entity that seemed to be unbroken yet everywhere jagged. Regarded by most people as unimaginable, his function not only refuted Ampère's "theorem" but drove the last nail into the coffin of geometric intuition as a trustworthy foundation for the calculus.

By all accounts, Weierstrass concocted his example in the 1860s and presented it to the Berlin Academy on July 18, 1872. As was his custom, he did not rush the discovery into print; it was first published by Paul du Bois-Reymond (1831–1889) in 1875.

Needless to say, so peculiar a function is far from elementary. In terms of technical complexity, it is probably the most demanding result in this book. But its counterintuitive nature, not to mention its historical significance, should make the effort worthwhile. Here we follow Weierstrass's argument but modify his notation and add a detail now and then for the sake of clarity.

We start with a lemma that Weierstrass would need later. He proved it with a trigonometric identity, but we present an argument using calculus.

Lemma: If $B > 0$, then $\left| \dfrac{\cos(A\pi + B\pi) - \cos(A\pi)}{B} \right| \le \pi.$

Proof: Let $h(x) = \cos(\pi x)$ over the interval $[A, A + B]$. By the mean value theorem, there is a point c between A and $A + B$ such that $\dfrac{h(A + B) - h(A)}{B} = h'(c)$. This amounts to $\dfrac{\cos(A\pi + B\pi) - \cos(A\pi)}{B} =$ $-\pi \sin(c\pi)$, and it follows that $\left| \dfrac{\cos(A\pi + B\pi) - \cos(A\pi)}{B} \right| =$ $|-\pi \sin(c\pi)| \le \pi \cdot 1 = \pi.$ Q.E.D.

We now introduce, in his own words, Weierstrass's famous counterexample.

Theorem: If $a \ge 3$ is an odd integer and if b is a constant strictly between 0 and 1 such that $ab > 1 + 3\pi/2$, then the function $f(x) =$ $\displaystyle\sum_{k=0}^{\infty} b^k \cos(\pi a^k x)$ is everywhere continuous and nowhere differentiable [8].

Dies kann z. B. folgendermassen geschehen.

Es sei x eine reelle Veränderliche, a eine ungrade ganze Zahl, b eine positive Constante, kleiner als 1, und

$$f(x) = \sum_{n=0}^{\infty} b^n \cos\left(a^n x \pi\right);$$

so ist $f(x)$ eine stetige Function, von der sich zeigen lässt, dass sie, sobald der Werth des Products ab eine gewisse Grenze übersteigt, an keiner Stelle einen bestimmten Differentialquotienten besitzt.

Weierstrass's pathological function (1872)

Proof: Obviously, he had done plenty of legwork before placing these strange restrictions upon a and b. To simplify the discussion, we shall let $a = 21$ and $b = 1/3$. These choices satisfy the stated conditions because $a \geq 3$ is an odd integer, b lies in $(0, 1)$, and $ab = 7 > 1 + 3\pi/2$. Consequently, our specific function will be

$$f(x) = \sum_{k=0}^{\infty} \frac{\cos(21^k \pi x)}{3^k} = \cos(\pi x) + \frac{\cos(21\pi x)}{3} + \frac{\cos(441\pi x)}{9} + \cdots. \tag{3}$$

To prove the continuity of f, we need only apply the M-test. Clearly

$$\left| \frac{\cos(21^k \pi x)}{3^k} \right| \leq \frac{1}{3^k} \quad \text{and} \quad \sum_{k=0}^{\infty} \frac{1}{3^k} \quad \text{converges to 3/2. Therefore, the series}$$

converges uniformly to f. Because each summand $\dfrac{\cos(21^k \pi x)}{3^k}$ is

continuous everywhere, so is f by theorem 1 above.

We seem to be halfway to showing that f is everywhere continuous and nowhere differentiable. However, proving the "nowhere differentiable" part is much, *much* more difficult. To this end, we begin by fixing a real number r. Our goal is to show that $f'(r)$ does not exist. Because r is arbitrary, this will establish that f is differentiable at no point whatever.

In following Weierstrass's logic, it will be helpful to assemble a number of observations about seemingly unrelated matters. Rest assured that each will play a role somewhere in his grand production.

First, Weierstrass noted that for each $m = 1, 2, 3, \ldots$, the real number $21^m r$ (like any real number) falls within half a unit of its nearest integer. Thus, for each whole number m, there exists an *integer* α_m such that

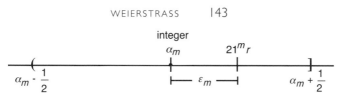

Figure 9.7

$\alpha_m - \dfrac{1}{2} < 21^m r \le \alpha_m + \dfrac{1}{2}$ (see figure 9.7). Letting $\varepsilon_m = 21^m r - \alpha_m$ be the associated gap, we see that

$$\alpha_m + \varepsilon_m = 21^m r. \tag{4}$$

Because $-\dfrac{1}{2} < \varepsilon_m \le \dfrac{1}{2}$, it follows that $0 < \dfrac{1/2}{21^m} \le \dfrac{1 - \varepsilon_m}{21^m} < \dfrac{3/2}{21^m}$.

For notational ease, we introduce $h_m = \dfrac{1 - \varepsilon_m}{21^m}$ and observe that

$$21^m h_m = 1 - \varepsilon_m \quad \text{and} \quad \frac{1}{h_m} > \frac{21^m}{3/2}. \tag{5}$$

Now, $0 < \dfrac{1/2}{21^m} \le h_m < \dfrac{3/2}{21^m}$ guarantees that $\displaystyle\lim_{m \to \infty} h_m = 0$ by the squeezing theorem. The sequence of positive terms $\{h_m\}$ will be decisive in establishing nondifferentiability.

At this point, we (temporarily) fix the integer m. As did Weierstrass, we use (3) and consider the differential quotient:

$$\frac{f(r + h_m) - f(r)}{h_m} = \frac{\displaystyle\sum_{k=0}^{\infty} \frac{\cos(21^k \pi [r + h_m])}{3^k} - \sum_{k=0}^{\infty} \frac{\cos(21^k \pi r)}{3^k}}{h_m}$$

$$= \sum_{k=0}^{m-1} \frac{\cos(21^k \pi r + 21^k \pi h_m) - \cos(21^k \pi r)}{3^k h_m}$$

$$+ \sum_{k=m}^{\infty} \frac{\cos(21^k \pi r + 21^k \pi h_m) - \cos(21^k \pi r)}{3^k h_m}. \tag{6}$$

Here, the infinite series has been broken into two parts. Weierstrass would consider the absolute value of each separately.

For the first series, we apply the lemma with $A = 21^k r$ and $B = 21^k h_m$ to bound each summand as follows:

$$\left| \frac{\cos(21^k \pi r + 21^k \pi h_m) - \cos(21^k \pi r)}{3^k h_m} \right|$$

$$= 7^k \left| \frac{\cos(21^k \pi r + 21^k \pi h_m) - \cos(21^k \pi r)}{21^k h_m} \right| \leq 7^k \pi.$$

Thus, by the triangle inequality, we have an upper bound for the first sum:

$$\left| \sum_{k=0}^{m-1} \frac{\cos(21^k \pi r + 21^k \pi h_m) - \cos(21^k \pi r)}{3^k h_m} \right|$$

$$\leq \sum_{k=0}^{m-1} \left| \frac{\cos(21^k \pi r + 21^k \pi h_m) - \cos(21^k \pi r)}{3^k h_m} \right|$$

$$\leq \sum_{k=0}^{m-1} 7^k \pi = \pi(1 + 7 + 49 + \cdots + 7^{m-1}) = \pi \left[\frac{7^m - 1}{6} \right] < \frac{\pi}{6}(7^m).$$

$$(7)$$

The second series in (6) presents a greater challenge. We approach the task by making four pertinent observations:

(A) If $k \geq m$, we see by (4) and (5) that

$$21^k \pi r + 21^k \pi h_m = 21^{k-m} \pi [21^m r + 21^m h_m]$$
$$= 21^{k-m} \pi [(\alpha_m + \varepsilon_m) + (1 - \varepsilon_m)]$$
$$= 21^{k-m} \pi [\alpha_m + 1].$$

But 21^{k-m} is an odd integer and α_m is an integer as well. Thus $21^{k-m} \pi [\alpha_m + 1]$ is an even or odd integer multiple of π depending on whether $\alpha_m + 1$ is even or odd. It follows that $\cos(21^k \pi r + 21^k \pi h_m) = \cos(21^{k-m} \pi [\alpha_m + 1]) = (-1)^{\alpha_m + 1}$.

(B) Again we stipulate that $k \geq m$ and apply (4) to get $21^k \pi r = 21^{k-m} \pi (21^m r) = 21^{k-m} \pi (\alpha_m + \varepsilon_m)$. By a familiar trig identity we have

$$\cos(21^k \pi r) = \cos(21^{k-m} \pi \alpha_m + 21^{k-m} \pi \varepsilon_m)$$
$$= \cos(21^{k-m} \pi \alpha_m) \cdot \cos(21^{k-m} \pi \varepsilon_m)$$
$$- \sin(21^{k-m} \pi \alpha_m) \cdot \sin(21^{k-m} \pi \varepsilon_m).$$

Here $21^{k-m}\pi\alpha_m$ is an integral multiple of π whose parity depends on α_m, and so

$$\cos(21^k\pi r) = (-1)^{\alpha_m} \cdot \cos(21^{k-m}\pi\varepsilon_m) - 0 \cdot \sin(21^{k-m}\pi\varepsilon_m)$$
$$= (-1)^{\alpha_m} \cdot \cos(21^{k-m}\pi\varepsilon_m).$$

(C) (An easy one) By the nature of cosine, $1 + \cos(21^{k-m}\pi\varepsilon_m) \geq 0$.

(D) Because $-\dfrac{1}{2} < \varepsilon_m \leq \dfrac{1}{2}$, we know that $-\dfrac{\pi}{2} < \pi\varepsilon_m \leq \dfrac{\pi}{2}$, and so $\cos(\pi\varepsilon_m) \geq 0$.

We now apply (A) and (B) to get a *lower* bound for the absolute value of the second series in (6):

$$\left| \sum_{k=m}^{\infty} \frac{\cos(21^k\pi r + 21^k\pi h_m) - \cos(21^k\pi r)}{3^k h_m} \right|$$

$$= \left| \sum_{k=m}^{\infty} \frac{(-1)^{\alpha_m+1} - (-1)^{\alpha_m} \cdot \cos(21^{k-m}\pi\varepsilon_m)}{3^k h_m} \right|$$

$$= \left| \sum_{k=m}^{\infty} \frac{(-1)^{\alpha_m+1}[1 + \cos(21^{k-m}\pi\varepsilon_m)]}{3^k h_m} \right|$$

$$= \left| \frac{(-1)^{\alpha_m+1}}{h_m} \right| \cdot \left| \sum_{k=m}^{\infty} \frac{1 + \cos(21^{k-m}\pi\varepsilon_m)}{3^k} \right|$$

$$= \frac{1}{h_m} \cdot \sum_{k=m}^{\infty} \frac{1 + \cos(21^{k-m}\pi\varepsilon_m)}{3^k},$$

because each term of the series is nonnegative by (C).

This sum of nonnegative terms is surely greater than its first term (where $k = m$), so by (D) and (5), we have

$$\left| \sum_{k=m}^{\infty} \frac{\cos(21^k\pi r + 21^k\pi h_m) - \cos(21^k\pi r)}{3^k h_m} \right|$$

$$\geq \frac{1}{h_m} \left[\frac{1 + \cos(\pi\varepsilon_m)}{3^m} \right] \geq \frac{1}{3^m h_m} > \frac{21^m}{3^m(3/2)} = \frac{2}{3}(7^m).$$

All of this has been a vast overture before the main performance. Weierstrass now derived the critical inequality, one that began with the result just proved and ended with a telling bound on the differential quotient:

$$\frac{2}{3}(7^m) < \left| \sum_{k=m}^{\infty} \frac{\cos(21^k \pi r + 21^k \pi h_m) - \cos(21^k \pi r)}{3^k h_m} \right|$$

$$= \left| \frac{f(r+h_m) - f(r)}{h_m} - \sum_{k=0}^{m-1} \frac{\cos(21^k \pi r + 21^k \pi h_m) - \cos(21^k \pi r)}{3^k h_m} \right|$$

by (6)

$$\leq \left| \frac{f(r+h_m) - f(r)}{h_m} \right| + \left| \sum_{k=0}^{m-1} \frac{\cos(21^k \pi r + 21^k \pi h_m) - \cos(21^k \pi r)}{3^k h_m} \right|$$

$$< \left| \frac{f(r+h_m) - f(r)}{h_m} \right| + \frac{\pi}{6}(7^m)$$

by (7).

From the first and last terms of this string of inequalities, we deduce that

$$\left| \frac{f(r+h_m) - f(r)}{h_m} \right| > \frac{2}{3}(7^m) - \frac{\pi}{6}(7^m) = \left[\frac{2}{3} - \frac{\pi}{6} \right] 7^m. \qquad (8)$$

Two features of expression (8) are critical. First, the quantity $\frac{2}{3} - \frac{\pi}{6} \approx 0.14307$ is a positive constant. Second, the inequality in (8) holds for our fixed, but arbitrary, whole number m. With this in mind, we now "unfix" m and take a limit:

$$\lim_{m \to \infty} \left| \frac{f(r+h_m) - f(r)}{h_m} \right| \geq \lim_{m \to \infty} \left[\frac{2}{3} - \frac{\pi}{6} \right] 7^m = \infty.$$

But we noted above that $h_m \to 0$ as $m \to \infty$. Therefore, $f'(r) = \lim_{h \to 0} \frac{f(r+h) - f(r)}{h}$ cannot exist as a finite quantity. In short (*short?*),

f is not differentiable at $x = r$. And because r was an unspecified real number, we have confirmed that Weierstrass's function, although everywhere continuous, is nowhere differentiable. Q.E.D.

Once the reader catches his or her breath, a number of reactions are likely. One is sheer amazement at Weierstrass's abilities. The talent involved in putting this proof together is quite extraordinary.

Another may be a sense of discomfort, for we have just verified that a continuous function may have no point of differentiability. Nowhere does its graph rise or fall smoothly. Nowhere does its graph have a tangent line. This is a function every point of which behaves like a sharp corner, yet which remains continuous throughout.

Would a picture of $y = f(x)$ be illuminating? Unfortunately, because f is an infinite series of functions, we must be content with graphing a partial sum. We do just that in figure 9.8 with a graph of the third partial sum

$$S_3(x) = \sum_{k=0}^{3} \frac{\cos(21^k \pi x)}{3^k} = \cos(\pi x) + \frac{\cos(21\pi x)}{3} + \frac{\cos(441\pi x)}{9}.$$

This reveals a large number of direction changes and some very steep rising and falling behavior, but no sharp angles. Indeed, any partial sum of Weierstrass's function, comprising finitely many cosines, is differentiable everywhere. No matter which partial sum we graph, we find not a single corner. Yet, when we pass to the limit to generate f itself, corners must

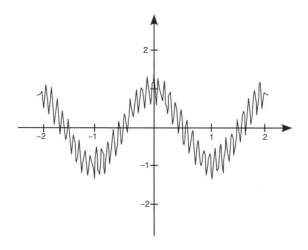

Figure 9.8

appear *everywhere*. Weierstrass's function lies somewhere beyond the intuition, far removed from geometrical diagrams that can be sketched on a blackboard. Yet its existence has been unquestionably established in the proof above.

A final reaction to this argument should be applause for its high standard of rigor. Like a maestro conducting a great orchestra, Weierstrass blended the fundamental definitions, the absolute values, and a host of inequalities into a coherent whole. Nothing was left to chance, nothing to intuition. For later generations of analysts, the ultimate compliment was to say that a proof exhibited "Weierstrassian rigor."

To be sure, not everyone was thrilled by a function so pathological. Some critics reacted against a mathematical world where inequalities trumped intuition. Charles Hermite, whom we met in the previous chapter, famously bemoaned the discovery in these words: "I turn away with fright and horror from this lamentable evil of functions that do not have derivatives" [9]. Henri Poincaré (1854–1912) called Weierstrass's example "an outrage against common sense" [10]. And the exasperated Emile Picard (1856–1941) wrote: "If Newton and Leibniz had thought that continuous functions do not necessarily have a derivative . . . the differential calculus would never have been invented" [11]. As though cast out of Eden, these mathematicians believed that paradise—in the form of an intuitive, geometric foundation for calculus—had been lost forever.

But Weierstrass's logic was ironclad. Short of abandoning the definitions of limit, continuity, and differentiability, or of denying analysts the right to introduce infinite processes, the critics were doomed. If something like a continuous, nowhere-differentiable function was intuitively troubling, then scholars needed to modify their intuitions rather than abandon their mathematics. Analytic rigor, advancing since Cauchy, reached a new pinnacle with Weierstrass. Like it or not, there was no turning back.

In a continuing ebb and flow, mathematicians develop grand theories and then find pertinent counterexamples to reveal the boundaries of their ideas. This juxtaposition of theory and counterexample is the logical engine by which mathematics progresses, for it is only by knowing how properties fail that we can understand how they work. And it is only by seeing how intuition misleads that we can truly appreciate the power of reason.

CHAPTER 10

Second Interlude

Our story has reached the year 1873, nearly a century after the passing of Euler and two after the creation of the calculus. By that date, the work of Cauchy, Riemann, and Weierstrass was sufficient to silence any latter-day Berkeley who might happen along. Was there anything left to do?

The answer, of course, is . . . "Of course." As mathematicians grappled with ideas like continuity and integrability, their very successes raised additional questions that were intriguing, troubling, or both. Weierstrass's pathological function was the most famous of many peculiar examples that suggested avenues for future research. Here we shall consider a few others, each of which will figure in the book's remaining chapters.

Our first is the so-called "ruler function," a simple but provocative example that appeared in a work of Johannes Karl Thomae (1840–1921) from 1875. He introduced it with this preamble: "Examples of integrable functions that are continuous or are discontinuous at individual points are plentiful, but it is important to identify integrable functions that are discontinuous infinitely often" [1].

His function was defined on the open interval (0, 1) by

$$r(x) = \begin{cases} 1/q & \text{if } x = p/q \text{ in lowest terms,} \\ 0 & \text{if } x \text{ is irrational.} \end{cases}$$

Thus, $r(1/5) = r(2/5) = r(4/10) = 1/5$, whereas $r(\pi/6) = r(1/\sqrt{2}) = 0$ Figure 10.1 displays the portion of its graph above $y = 1/7$; below this, the scattered points become impossibly dense. The graph suggests the vertical markings on a ruler—hence the name.

With the ε-δ definition from the previous chapter, it is easy to prove the following lemma.

Lemma: If a is any point in (0, 1), then $\lim_{x \to a} r(x) = 0$.

Proof: For $\varepsilon > 0$, we chose a whole number N with $1/N < \varepsilon$. The proof rests upon the observation that in (0, 1) there are only finitely many

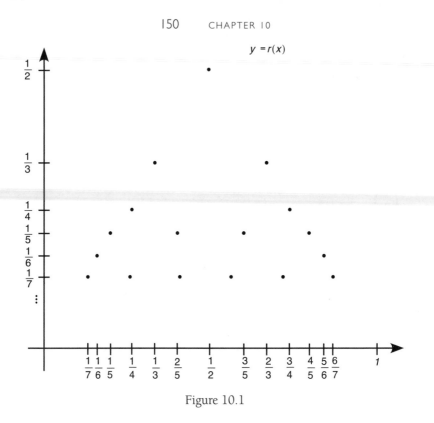

Figure 10.1

rationals in lowest terms whose denominators are N or smaller. For example, the only such fractions with denominators 5 or smaller are 1/2, 1/3, 2/3, 1/4, 3/4, 1/5, 2/5, 3/5, and 4/5. Because this collection is finite, we can find a positive number δ small enough that the interval $(a - \delta, a + \delta)$ lies within $(0, 1)$ and contains *none* of these fractions, except possibly a itself. We now choose any x with $0 < |x - a| < \delta$ and consider two cases. If $x = p/q$ is a rational in lowest terms, then $|r(x) - 0| = |r(p/q)| = 1/q < 1/N < \varepsilon$ because q must be greater than N if $p/q \neq a$ is in $(a - \delta, a + \delta)$. Alternately, if x is irrational, then $|r(x) - 0| = 0 < \varepsilon$ as well. In either case, for $\varepsilon > 0$, we have found a $\delta > 0$ so that, if $0 < |x - a| < \delta$, then $|r(x) - 0| < \varepsilon$. By definition, $\lim\limits_{x \to a} r(x) = 0$.

Q.E.D.

With the lemma behind us, we can demonstrate the ruler function's most astonishing property: it is continuous at each irrational in $(0, 1)$ yet discontinuous at each rational in $(0, 1)$. This follows immediately because, if a is irrational, then $r(a) = 0 = \lim\limits_{x \to a} r(x)$ by the lemma—precisely Cauchy's definition of continuity at a. On the other hand, if $a = p/q$ is a rational in lowest terms, then

$$r(a) = r(p/q) = 1/q \neq 0 = \lim_{x \to a} r(x),$$

and so the ruler function is discontinuous at a.

This presents us with a bizarre situation: the function is continuous (which our increasingly unreliable intuition regards as "unbroken") at irrational points but discontinuous ("broken") at rational ones. Most of us find it impossible to envision how the continuity/discontinuity points can be so intertwined. But the mathematics above is unambiguous.

It will be helpful to extend the domain of the ruler function from $(0, 1)$ to all real numbers. This is done by letting our new function take the value 1 at each integer and putting copies of r on each subinterval $(1, 2)$, $(2, 3)$, and so on. More precisely, we define the extended ruler function R by

$$R(x) = \begin{cases} 1 & \text{if } x \text{ is an integer,} \\ r(x - n) & \text{if } n < x < n+1 \text{ for some integer } n \geq 0, \\ r(x + n + 1) & \text{if } -(n+1) < x < -n \text{ for some integer } n \geq 0. \end{cases}$$

As above, we have $\lim_{x \to a} R(x) = 0$ for any real number a, and so R is continuous at each irrational and discontinuous at each rational.

The ruler function raises a natural question: "How can we flip-flop roles and create a function that is continuous at each rational and discontinuous at each irrational?" Although simple to state, this has a profound, and profoundly intriguing, answer. It will be the main topic in our upcoming chapter on Vito Volterra.

The ruler function R is also remarkable because, its infinitude of discontinuities notwithstanding, it is integrable over $[0, 1]$. That, of course, is the essence of Thomae's preamble above. To prove it, we use Riemann's integrability condition from chapter 7.

Begin with a value of $d > 0$ and a fixed oscillation $\sigma > 0$. We then choose a whole number N such that $1/N < \sigma$. As in the argument above, we know that $[0,1]$ contains only finitely many rationals in lowest terms for which $R(p/q) \geq 1/N$, namely those with denominators no greater than N. We let M be the number of such rationals and partition $[0,1]$ so that each of these lies within a subinterval of width $d/2M$. These will be what we called the Type A subintervals, that is, those where the function oscillates more than σ. Using Riemann's terminology, we have

$$s(\sigma) = \sum_{\text{Type A}} \delta_k = \sum_{\text{Type A}} \frac{d}{2M} \leq M\left(\frac{d}{2M}\right) = \frac{d}{2},$$

so that $s(\sigma) \to 0$ as $d \to 0$. This is exactly what Riemann needed to establish integrability. In other words, $\int_0^1 R(x)dx$ exists. Further, knowing that the integral exists, we can easily show that $\int_0^1 R(x)dx = 0$.

It should be plain that the ruler function plays the same role as Riemann's pathological function from chapter 7. Both are discontinuous infinitely often, yet both are integrable. The major difference between them is the ruler function's relative simplicity, and, under the circumstances, a little simplicity is nothing to be sneered at.

There is an intriguing question raised by these examples. We recall that Dirichlet's function was *everywhere* discontinuous and not Riemann integrable. By contrast, the ruler function is discontinuous only on the rationals. This, to be sure, is awfully discontinuous, but the function still possesses enough continuity to allow it to be integrated. With such evidence, mathematicians conjectured that a Riemann-integrable function could be discontinuous, but not *too* discontinuous. Coming to grips with the continuity/integrability issue would occupy analysts for the remainder of the nineteenth century. As we shall see in the book's final chapter, this matter was addressed, and ultimately resolved, by Henri Lebesgue in 1904.

Our next three examples are interrelated and so can be treated together. Like the ruler function, these are fixtures in most analysis textbooks because of their surprising properties.

First, we define $S(x) = \begin{cases} \cos(1/x) & \text{if } x \neq 0, \\ 0 & \text{if } x = 0, \end{cases}$ and graph it in figure 10.2.

As x approaches zero, its reciprocal $1/x$ grows without bound, causing $\cos(1/x)$ to gyrate from -1 to 1 and back again infinitely often in any neighborhood of the origin. To say that S oscillates wildly is an understatement.

We show that $\lim_{x \to 0} S(x)$ does not exist by introducing the sequence $\{1/k\pi\}$ for $k = 1, 2, 3, \ldots$ and looking at the corresponding points on the graph. As indicated in figure 10.2, we are alternately selecting the crests and valleys of our function. That is, $\lim_{k \to \infty}(1/k\pi) = 0$, but

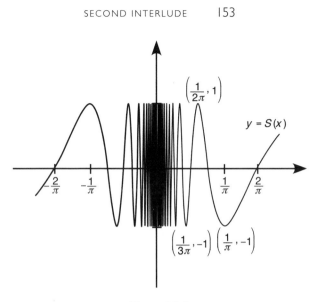

Figure 10.2

$\lim\limits_{k\to\infty} S(1/k\pi) = \lim\limits_{k\to\infty}[\cos(k\pi)] = \lim\limits_{k\to\infty}(-1)^k$. Because this last limit does not exist, neither does $\lim\limits_{x\to 0} S(x)$, which in turn means that S is discontinuous at $x = 0$.

A related function is $T(x) = \begin{cases} x\sin(1/x) & \text{if } x \neq 0, \\ 0 & \text{if } x = 0, \end{cases}$ which is graphed in figure 10.3. Because of the multiplier x, the infinitely many oscillations of T damp out as we approach the origin.

At any nonzero point, T is the product of the continuous functions $y = x$ and $y = \sin(1/x)$ and so is itself continuous. Because $-|x| \leq x\sin(1/x) \leq |x|$ and $\lim\limits_{x\to 0}(-|x|) = 0 = \lim\limits_{x\to 0}|x|$, the squeezing theorem guarantees that $\lim\limits_{x\to 0} T(x) = 0 = T(0)$, so T is continuous at $x = 0$ as well. In short, T is an everywhere-continuous function. It is often cited as an example to show that "continuous" is not the same as "able to be drawn without lifting the pencil." The latter may be a useful characterization in the first calculus course, but graphing $y = T(x)$ in a neighborhood of the origin is impossible with all those ups and downs.

Finally, we consider the most provocative member of our trio:

$$U(x) = \begin{cases} x^2\sin(1/x) & \text{if } x \neq 0, \\ 0 & \text{if } x = 0. \end{cases}$$

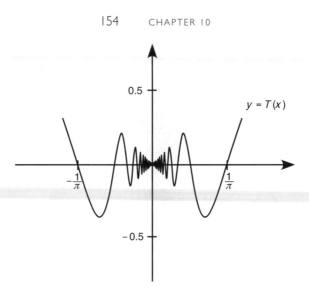

Figure 10.3

Here the quadratic coefficient accelerates the damping of the curve near the origin. Because $U(x) = x\,T(x)$ and both factors are everywhere continuous, so is U.

This time the troubling issue involves differentiability. At any $x \neq 0$, the function is certainly differentiable, and the rules of calculus show that $U'(x) = 2x\,\sin(1/x) - \cos(1/x)$. At $x = 0$ the function is differentiable as well because

$$U'(0) = \lim_{x \to 0} \frac{U(x) - U(0)}{x - 0} = \lim_{x \to 0} \frac{x^2 \sin(1/x)}{x} = \lim_{x \to 0} [x \sin(1/x)] = 0,$$

where the final limit employs the same "squeeze" we just saw. So, in spite of its being infinitely wobbly near the origin, the function U has a horizontal tangent there.

We have proved that U is everywhere differentiable with

$$U'(x) = \begin{cases} 2x\sin(1/x) - \cos(1/x) & \text{if } x \neq 0, \\ 0 & \text{if } x = 0. \end{cases}$$

Alas, this derivative is not a continuous function, for we again consider the sequence $\{1/k\pi\}$ and note that

$$\lim_{k \to \infty} U'\left(\frac{1}{k\pi}\right) = \lim_{k \to \infty} \left[\frac{2}{k\pi} \sin(k\pi) - \cos(k\pi) \right] = \lim_{k \to \infty} [0 - (-1)^k],$$

which does not exist. Thus, $\lim_{x \to 0} U'(x)$ cannot exist and so U' is discontinuous at $x = 0$. In short, U is a differentiable function with a *discontinuous* derivative.

This brings to mind the famous theorem that a differentiable function is continuous. It would be natural to propose the following modification: "The *derivative* of a differentiable function must be continuous." The example of U, however, shows that such a modification is wrong.

These examples also muddy the relationship between continuity and the intermediate value theorem. As we saw, Cauchy proved that a continuous function must take all values between any two that it assumes. This geometrically self-evident fact might appear to be the very essence of continuity, and one could surmise that a function is continuous if and only if it possesses the intermediate value property over every interval of its domain.

Again, this assumption turns out to be erroneous. As a counterexample, consider S from above. We have seen that S is discontinuous at the origin, but we claim that it has the intermediate value property over every interval.

To prove this, suppose $S(a) \le r \le S(b)$ for $a < b$. By the nature of the cosine, we know that $-1 \le r \le 1$. We now consider cases:

If $0 < a < b$ or if $a < b < 0$, then S is continuous throughout $[a, b]$ and so, for some c in (a, b), we have $S(c) = r$ by the intermediate value theorem.

On the other hand, if $a < 0 < b$, we can fix a whole number N with $N > \dfrac{1}{2\pi b}$ Then $a < 0 < \dfrac{1}{(2N+1)\pi} < \dfrac{1}{2N\pi} < b$, and as x runs between the positive numbers $\dfrac{1}{(2N+1)\pi}$ and $\dfrac{1}{2N\pi}$, the value of $1/x$ runs between $2N\pi$ and $(2N+1)\pi$. In the process, $S(x) = \cos(1/x)$ goes continuously from $\cos(2N\pi) = 1$ to $\cos[(2N+1)\pi] = -1$. By the intermediate value theorem, there must be a c between $\dfrac{1}{(2N+1)\pi}$ and $\dfrac{1}{2N\pi}$ (and consequently between a and b) for which $S(c) = r$. The claim is thus proved.

In summary, our examples have shown that the derivative of a differentiable function need not be continuous and that a function possessing the intermediate value property need not be continuous either. These may seem odd, but there is one last surprise in store.

It was discovered by Gaston Darboux (1842–1917), a French mathematician who is known for a pair of contributions to analysis. First, he simplified the development of the Riemann integral so as to achieve the same end in a much less cumbersome fashion. Today's textbooks, when they introduce the integral, tend to use Darboux's elegant treatment instead of Riemann's original.

But it is the other contribution we address here. In what is now called "Darboux's theorem," he proved that derivatives, although not necessarily continuous, *must* possess the intermediate value property. The argument rests upon two results that appear in any introductory analysis text: one is that a continuous function takes a minimum value on a closed, bounded interval [a, b], and the other is that $g'(c) = 0$ if g is a differentiable function with a minimum at $x = c$.

Darboux's Theorem: If f is differentiable on [a, b] and if r is any number for which $f'(a) < r < f'(b)$, then there exists a c in (a, b) such that $f'(c) = r$.

Proof: To begin, we introduce a new function $g(x) = f(x) - rx$. Because f is differentiable, it is continuous, and rx is continuous as well, so g is continuous on [a, b]. Thus, for some point c in [a, b], g takes a minimum value.

The differentiability of f implies that g is also differentiable, with $g'(x) = f'(x) - r$. Clearly, $g'(a) = f'(a) - r < 0$ and $g'(b) = f'(b) - r > 0$. Yet we know that $g'(c) = 0$ by the second result above. It follows that c is neither a nor b and so must lie somewhere between. For this c in (a, b), we have

$$0 = g'(c) = f'(c) - r, \text{ or simply } f'(c) = r.$$

Thus f' assumes the intermediate value r, as required.   Q.E.D.

The reader will recall that in Cauchy's proof of the mean value theorem, he assumed his derivative was continuous in order to conclude that it took intermediate values. We now see that Cauchy could have discarded his assumption without discarding his conclusion. It also follows that a function lacking the intermediate value property, for instance, Dirichlet's function, cannot be the derivative of *anything*.

Darboux showed that derivatives share with continuous functions the property of taking intermediate values. And this suggests another question: "How discontinuous can a derivative really be?" As we see in the book's next-to-last chapter, René Baire provided an answer in 1899.

If derivatives were troubling, integrals were more so. We noted previously that, even when the sequence $\{f_k\}$ converges pointwise, we cannot generally conclude that

$$\lim_{k \to \infty} \left[\int_a^b f_k(x)dx \right] = \int_a^b \left[\lim_{k \to \infty} f_k(x) \right] dx. \tag{1}$$

Weierstrass showed that uniform convergence is sufficient to guarantee the interchange of limits and integrals, but it turns out not to be necessary. That is, examples $\{f_k\}$ were found that converged pointwise but not uniformly and yet for which (1) holds. Perhaps mathematicians had overlooked some intermediate condition, not so restrictive as uniform convergence, that would allow the much-desired interchange.

Or—and this at first seemed a very unlikely "or"—perhaps Riemann's definition of the integral was at fault. In treating integration as he did, Riemann may have taken the wrong path, one that required special conditions in order for (1) to hold. If so, his integral could be regarded as defective.

On the face of it, this sounded like heresy, for Riemann's integral had become a pillar of mathematical analysis. Darboux described it as a creation "of which only the greatest minds are capable" [2]. And Paul du Bois-Reymond stated his belief that Riemann's definition could not be improved upon, for it extended the concept of integrability to its outermost limits [3]. Yet, as we shall see, this and other shortcomings motivated research aimed at defining the integral more broadly. The result would be Lebesgue's theory of integration from the turn of the twentieth century.

To summarize, the functions above raised such questions as:

- Can we construct a function continuous at each rational and discontinuous at each irrational?
- How discontinuous can a Riemann integrable function be?
- How discontinuous can a derivative be?
- How, if at all, can we correct the deficiencies in the Riemann integral?

Although not an exhaustive list, these were critical issues confronting mathematical analysis as the nineteenth century entered its final quarter. By their very nature, such questions could hardly have been *asked*, let alone answered, before the contributions of Cauchy, Riemann, and Weierstrass. As the challenges grew ever more sophisticated, their resolutions would require increasingly careful reasoning. In the remainder of the book, we shall indicate how each of these four questions was answered.

Our first stop, however, will be an 1874 paper by Georg Cantor, the genius who gave birth to set theory and applied his ideas to re-prove the existence of transcendentals. His achievement illustrates as well as anything the benefits of thinking anew about matters long regarded as settled.

CHAPTER II

Cantor

Georg Cantor

The essence of mathematics lies in its freedom" [1]. So wrote Georg Cantor (1845–1918) in 1883. Few mathematicians so thoroughly embraced this principle and few so radically changed the nature of the subject. Joseph Dauben, in his study of Cantor's works, described him as "one of the most imaginative and controversial figures in the history of mathematics" [2]. The present chapter should demonstrate why this assessment is valid.

Cantor came from a line of musicians, and it is possible to see in him tendencies more often associated with the romantic artist than with the pragmatic technician. His research eventually carried him beyond mathematics to the borders of metaphysics and theology. He raised many an eyebrow with claims that Francis Bacon had written the Shakespearean canon and that his own theory of the infinite proved the existence of God. As an uncompromising advocate of such beliefs, Cantor had a way of alienating friend and foe alike.

Meanwhile, his life was troubled. He suffered bouts of severe depression, almost certainly a bipolar disorder whose recurrences robbed him of the "mental freshness" he so coveted [3]. Time and again Cantor was sent to what were called neuropathic hospitals to endure whatever treatment they could offer. In 1918 he died in a psychiatric institution after a life with more than its share of unhappiness.

None of this detracts from Cantor's mathematical triumph. For all of his misfortune, Georg Cantor revolutionized the subject whose freedom he so loved.

THE COMPLETENESS PROPERTY

As a young man, Cantor had studied with Weierstrass at the University of Berlin. There he wrote an 1867 dissertation on number theory, a field very different from that for which he would become known. His research led him to Fourier series and eventually to the foundations of analysis.

As we have seen, developments in the nineteenth century placed calculus squarely upon the foundation of limits. It had become clear that limits, in turn, rested upon properties of the real number system, foremost among which is what we now call *completeness*. Today's students may encounter completeness in different but logically equivalent forms, such as:

C1. Any nondecreasing sequence that is bounded above converges to some real number.

C2. Any Cauchy sequence has a limit.

C3. Any nonempty set of real numbers with an upper bound has a least upper bound.

Readers in need of a quick refresher are reminded that $\{x_k\}$ is a *Cauchy sequence* if, for every $\varepsilon > 0$, there exists a whole number N such that, if m and n are whole numbers greater than or equal to N, then $|x_m - x_n| < \varepsilon$. In words, a Cauchy sequence is one whose terms get and stay close to one another. This idea put in a brief appearance in chapter 6.

Likewise, M is said to be an *upper bound* of a nonempty set A if $a \leq M$ for all elements a in A, and λ is a *least* upper bound, or *supremum*, of A if (1) λ is an upper bound of A and (2) if M is any upper bound of A, then $\lambda \leq M$. These concepts appear in any modern analysis text.

There is one other version of completeness, cast in terms of nested intervals, that will play an important role in the next few chapters. Again, we need a few definitions to clarify what is going on.

A closed interval $[a, b]$ is nested within $[A, B]$ if the former is a subset of the latter. This amounts to nothing more than the condition that $A \leq a \leq b \leq B$. Suppose further that we have a sequence of closed, bounded intervals, each nested within its predecessor, as in $[a_1, b_1] \supseteq [a_2, b_2] \supseteq [a_3, b_3] \supseteq \cdots \supseteq [a_k, b_k] \supseteq \cdots$. Such a sequence is said to be *descending*. With this we can introduce another version of completeness:

C4. Any descending sequence of closed, bounded intervals has a point that belongs to each of the intervals.

It is worth recalling why the intervals in question must be both closed and bounded. The descending sequence of closed (but not bounded) intervals

$$[1, \infty) \supseteq [2, \infty) \supseteq [3, \infty) \supseteq \cdots \supseteq [k, \infty) \supseteq \cdots$$

has no point common to all of them, and the descending sequence of bounded (but not closed) intervals

$$(0, 1) \supseteq (0, 1/2) \supseteq (0, 1/3) \supseteq \cdots \supseteq (0, 1/k) \supseteq \cdots$$

likewise has an empty intersection (to use set-theoretic terminology). Although our nineteenth century predecessors often neglected such distinctions, we shall arrange for our intervals to be both closed and bounded before applying C4.

Each of these four incarnations of completeness guarantees that some real number *exists*, be it the limit to which a sequence converges, or the least upper bound that a set possesses, or a point common to each of a collection of nested intervals. As mathematicians probed the logical foundations of calculus, they realized that such existence was often sufficient for their theoretical purposes. Rather than identify a real number explicitly, it may be enough to know that a number is out there somewhere. Completeness provides that assurance.

One might ask: if the completeness property is so important, how do we prove it? The answer required mathematicians to understand the real number system itself. From the whole numbers, it is a straightforward task to define the integers (positive, negative, and zero) and from there to define the rationals. But can we create the real numbers from more elementary systems, just as the rationals were defined in terms of the integers?

Affirmative answers to this question came from Cantor and, independently, from his friend Richard Dedekind (1831–1916). Cantor's construction of the reals was based on equivalence classes of Cauchy sequences of rational numbers. Dedekind's approach employed partitions of the rationals into disjoint classes, the so-called "Dedekind cuts." A thorough discussion of these matters would carry us far afield, for constructing the real numbers from the rationals is a bit esoteric for this book and, truth be told, a bit esoteric for most analysis courses. Nonetheless, Cantor and Dedekind did it successfully and then used their ideas to prove the completeness property as a theorem in their newly created realm.

This achievement can be seen as the final step in the separation of calculus from geometry. Dedekind and Cantor had gone back to the arithmetic basics—the whole numbers—from which the reals, then the completeness property, and eventually all of analysis could be developed. Their achievement received the apt but nearly unpronounceable moniker: "the arithmetization of analysis."

THE NONDENUMERABILITY OF INTERVALS

It is not for defining the real numbers that Cantor has been chosen to headline this chapter. Rather it is for his 1874 paper, *"Über eine Eigenschaft des Inbegriffes aller reellen algebraischen Zahlen"* (On a Property of the Totality of All Real Algebraic Numbers) [4]. This was a landmark in the history of mathematics, one that demonstrated, in Dauben's words, "[Cantor's] gift for posing incisive questions and for sometimes finding unexpected, even unorthodox answers" [5].

Oddly, the significance of the paper was obscured by its title, for the result about algebraic numbers was but a corollary, albeit a most interesting one, to the paper's truly revolutionary idea. That idea, simply stated, is that a sequence cannot exhaust an open interval of real numbers. As we shall see, Cantor's argument involved the completeness property, thus placing it properly in the domain of real analysis.

Theorem: If $\{x_k\}$ is a sequence of distinct real numbers, then any open, bounded interval (α, β) of real numbers contains a point not included among the $\{x_k\}$.

Proof: Cantor began with an interval (α, β) and considered the sequence in consecutive order: $x_1, x_2, x_3, x_4, \ldots$. If none or just one of these terms lies among the infinitude of real numbers in (α, β), then the proposition is trivially true.

Suppose, instead, that the interval contains at least two sequence points. We then identify the first two terms, by which we mean those with the two smallest subscripts, that fall within (α, β). We denote the smaller of these by A_1 and the larger by B_1. This step is illustrated in figure 11.1. Note that the initial few terms of the sequence fall outside of (α, β) but that x_4 and x_7 fall within it. By our definition, $A_1 = x_7$ (the smaller) and $B_1 = x_4$ (the greater).

We make two simple but important observations:

1. $\alpha < A_1 < B_1 < \beta$, and
2. if a sequence term x_k falls within the open interval (A_1, B_1), then $k \geq 3$.

The second of these recognizes that at least two sequence terms are used up in identifying A_1 and B_1, so any term lying strictly between A_1 and B_1 must have subscript $k = 3$ or greater. In figure 11.1, the next such candidate would be x_8.

Cantor then examined (A_1, B_1) and considered the same pair of cases: either this open interval contains none or just one of the terms of $\{x_k\}$ or it contains at least two of them. In the first case the theorem is true, for there are infinitely many other points in (A_1, B_1), and thus in (α, β), that do not belong to the sequence $\{x_k\}$. In the second case, Cantor repeated the earlier process by choosing the next two terms of the sequence, that is, those with the smallest subscripts, that fall within (A_1, B_1). He labeled the smaller of these A_2, and the larger B_2. If we look at figure 11.2 (which includes more terms of the sequence than did figure 11.1), we see that $A_2 = x_{10}$ and that $B_2 = x_{11}$.

Here again it is clear that

1. $\alpha < A_1 < A_2 < B_2 < B_1 < \beta$, and
2. if x_k falls within the open interval (A_2, B_2), then $k \geq 5$.

As before, the latter observation follows because at least four terms of the sequence $\{x_k\}$ must have been consumed in finding A_1, B_1, A_2, and B_2.

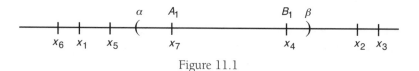

Figure 11.1

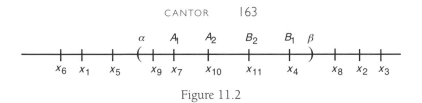

Figure 11.2

Cantor continued in this manner. If at any step there were one or fewer sequence terms remaining within the open subinterval, he could immediately find a point—indeed infinitely many of them—belonging to (α, β) but not to the sequence $\{x_k\}$. The only potential difficulty arose if the process never terminated, thereby generating a pair of infinite sequences $\{A_r\}$ and $\{B_r\}$ such that

1. $\alpha < A_1 < A_2 < A_3 < \cdots < A_r < \cdots < B_r < \cdots < B_3 < B_2 < B_1 < \beta$, and
2. if x_k falls within the open interval (A_r, B_r), then $k \geq 2r + 1$.

We then have a descending sequence of closed and bounded intervals $[A_1, B_1] \supseteq [A_2, B_2] \supseteq [A_3, B_3] \supseteq \cdots$, each nested within its predecessor. By the completeness property (C4), there is at least one point common to all of the $[A_r, B_r]$. That is, there exists a point c belonging to $[A_r, B_r]$ for *all* $r \geq 1$. To finish the proof, we need only establish that c lies in (α, β) but is not a term of the sequence $\{x_k\}$.

The first observation is immediate, for c is in $[A_1, B_1] \subset (\alpha, \beta)$ and so c indeed falls within the original open interval (α, β).

Could c appear as a term of the sequence $\{x_k\}$? If so, then $c = x_N$ for some subscript N. Because c lies in all of the closed intervals, it lies in $[A_{N+1}, B_{N+1}]$, and thus

$$A_N < A_{N+1} \leq c \leq B_{N+1} < B_N.$$

It follows that $c = x_N$ lies in the open interval (A_N, B_N), and so, according to (2) above, $N \geq 2N + 1$. This, of course, is absurd. We conclude that c can be none of the terms in the sequence $\{x_k\}$.

To summarize, Cantor had demonstrated that in (α, β) there is a point not appearing in the original sequence $\{x_k\}$. The existence of such a point was the object of the proof.   Q.E.D.

Today, this theorem is usually preceded by a bit of terminology. We define a set to be *denumerable* if it can be put into a one-to-one correspondence with the set of whole numbers. Sequences are trivially denumerable, with the required correspondence appearing as the subscripts. An infinite set that cannot be put into a one-to-one correspondence with the whole numbers is said to be *nondenumerable*. We then characterize the result above as proving that any open interval of real numbers is nondenumerable.

The evolution of Cantor's thinking on this matter is interesting. Through the early 1870s, he had pondered the fundamental properties of the real numbers, trying to isolate exactly what set them apart from the rationals. Obviously, completeness was a key distinction that somehow embodied what was meant by "the continuum" of the reals.

But Cantor began to suspect there was a difference in the *abundance* of numbers in these two sets—what we now call their "cardinality"—and in November of 1873 shared with Dedekind his doubts that the whole numbers could be matched in a one-to-one fashion with the real numbers. Implicitly this meant that, although both collections were infinite, the reals were more so.

Try as he might, Cantor could not prove his hunch. He wrote Dedekind, in some frustration, "as much as I am inclined to the opinion that [the whole numbers] and [the real numbers] permit no such unique correspondence, I cannot find the reason" [6]. A month later, Cantor had a breakthrough. As a Christmas gift to Dedekind, he sent a draft of his proof and, after receiving suggestions from the latter, cleaned it up and published what we saw above. Persistence had paid off.

Readers who know Cantor's "diagonalization" proof of nondenumerability may be surprised to see that his 1874 reasoning was wholly different. The diagonal argument, which Cantor described as a "much simpler demonstration," appeared in an 1891 paper [7]. In contrast to the 1874 proof, which, as we have seen, invoked the completeness property, diagonalization was applicable to situations where completeness was irrelevant, far from the constraints of analysis proper.

Although the later argument is more familiar, the earlier one represents the historic beginning and so has been included here. We stress again that Cantor's original proof did not use terms like denumerability nor raise specific questions about infinite cardinalities. All this would come later. In 1874, he simply showed that a sequence cannot exhaust an open interval.

But why should anyone care? It was a good question, and Cantor had a spectacular answer.

THE EXISTENCE OF TRANSCENDENTALS, REVISITED

We recall that Cantor's paper was titled, "On a Property of the Totality of All Real Algebraic Numbers." To this point, algebraic numbers have yet to be mentioned, nor have we said anything about the "property" of these numbers to which the title refers. The time has come to address those omissions.

As we saw, a real number is algebraic if it is the solution to a polynomial equation with integer coefficients. There are infinitely many of these (for instance, any rational number), and it was no easy matter for Liouville to find a number that lay outside the algebraic realm.

Cantor, upon considering the matter, claimed that it was possible to list the algebraic numbers in a sequence. At first glance, this may seem preposterous. It would require him to generate a sequence with the twin properties that (1) every term was an algebraic number and (2) every algebraic number was somewhere in the sequence. A clever eye would be necessary to do this in an orderly and exhaustive fashion, but Cantor was nothing if not clever. He began by introducing a new idea.

Definition: If $P(x) = ax^n + bx^{n-1} + cx^{n-2} + \cdots + gx + h$ is an nth-degree polynomial with integer coefficients, we define its *height* by $(n-1) + |a| + |b| + |c| + \cdots + |h|$.

For instance, the height of $P(x) = 2x^3 - 4x^2 + 5$ is $(3-1) + 2 + 4 + 5 = 13$ and that of $Q(x) = x^6 - 6x^4 - 10x^3 + 12x^2 - 60x + 17$ is $(6-1) + 1 + 6 + 10 + 12 + 60 + 17 = 111$.

Clearly the height of a polynomial with integer coefficients will itself be a whole number. Further, any algebraic number has a minimal-degree polynomial whose coefficients we can assume to have no common divisor other than 1. These conventions simplify the task at hand.

Cantor in turn collected all algebraic numbers that arise from polynomials of height 1, then those that arise from polynomials of height 2, then of height 3, and so on. This was the key to arranging algebraic numbers into an infinite sequence, here denoted by $\{a_k\}$.

To see the process in action, we observe that the only polynomial with integer coefficients of height 1 is $P(x) = 1 \cdot x^1 = x$. The solution to the associated equation $P(x) = 0$ is the first algebraic number, namely $a_1 = 0$.

There are four polynomials with height 2:

$$P_1(x) = x^2, \, P_2(x) = 2x, \, P_3(x) = x + 1, \, P_4(x) = x - 1.$$

Setting the first and second equal to zero yields the solution $x = 0$, which we do not count again. Setting $P_3(x) = 0$ gives $a_2 = -1$ and $P_4(x) = 0$ gives $a_3 = 1$.

We continue. There are eleven polynomials of height 3:

$$P_1(x) = x^3, \, P_2(x) = 2x^2, \, P_3(x) = x^2 + 1, \, P_4(x) = x^2 - 1,$$

$$P_5(x) = x^2 + x, \, P_6(x) = x^2 - x, \, P_7(x) = 3x, \, P_8(x) = 2x + 1,$$

$$P_9(x) = 2x - 1, \, P_{10}(x) = x + 2, \, P_{11}(x) = x - 2.$$

Upon setting these equal to zero, we get four new algebraic numbers:

$$a_4 = -\frac{1}{2}, \, a_5 = \frac{1}{2}, \, a_6 = -2, \text{ and } a_7 = 2.$$

As his title indicated, Cantor was restricting his attention to *real* algebraic numbers, so $0 = P_3(x) = x^2 + 1$ added nothing to the collection.

And on we go. There are twenty-eight polynomials of height 4, and from these we harvest a dozen additional algebraic numbers, some of which are irrational. For instance, the polynomial $P(x) = x^2 + x - 1$ is of height 4 and contributes $\dfrac{-1 + \sqrt{5}}{2}$ and $\dfrac{-1 - \sqrt{5}}{2}$.

As the heights increase, more and more algebraic numbers appear. Conversely, any specific algebraic number must arise from *some* polynomial with integer coefficients, and this polynomial, in turn, has a height. For instance, the algebraic number $\sqrt{2} + \sqrt[3]{5}$, which we encountered in chapter 8, is a solution to the polynomial equation $x^6 - 6x^4 - 10x^3 + 12x^2 - 60x + 17 = 0$ with height 111.

A few simple observations allowed Cantor to wrap up his argument:

- For a given height, there are only finitely many polynomials with integer coefficients.
- Each such polynomial can generate only finitely many new algebraic numbers (because an nth-degree polynomial equation can have no more than n solutions).
- Hence, for each height there can be only finitely many new algebraic numbers.

This means that, upon "entering" a given height in our quest for algebraic numbers, we must emerge from that height after finitely many steps. We cannot get "stuck" in a height trying to list an infinitude of new algebraic numbers.

Consequently, the number $\sqrt{2} + \sqrt[3]{5}$ with its polynomial of height 111 has to show up somewhere in our sequence $\{a_k\}$. It will take a while, but the process must, after finitely many steps, bring us to height 111, and then, as we run through the polynomials of this height, we reach $x^6 - 6x^4 - 10x^3 + 12x^2 - 60x + 17$ after finitely many more. This will determine the position of $\sqrt{2} + \sqrt[3]{5}$ in the sequence $\{a_k\}$. The same can be said of any real algebraic number. So, the "property" of the algebraic numbers mentioned in Cantor's title is, in modern parlance, its denumerability.

Now he combined his two results: first, that a sequence cannot exhaust an interval and, second, that the algebraic numbers form a sequence. Individually, these are interesting. Together, they allowed him to conclude that the algebraic numbers cannot account for all points on an open interval. Consequently, within any (α, β), there must lie a transcendental.

Or, to put it directly, transcendental numbers *exist*.

Of course, this was what Liouville had demonstrated a few decades earlier when he showed that $\sum_{k=1}^{\infty} \frac{1}{10^{k!}} = \frac{1}{10} + \frac{1}{10^2} + \frac{1}{10^6} + \frac{1}{10^{24}} + \frac{1}{10^{120}} + \cdots$ was transcendental. To prove the existence of transcendental numbers, he went out and found one.

Cantor reached the same end by very different means. Early in his 1874 paper, he had promised "a new proof of the theorem first demonstrated by Liouville," and he certainly delivered [8]. But his argument, as we have seen, contained no example of a specific transcendental. It was strikingly nonexplicit.

To contrast the two approaches, we offer the analogy of finding a needle in a haystack. We envision Liouville, industrious to a fault, putting on his old clothes, hiking out to the field, and rooting around in the hay under a broiling sun. Hours later, drenched with perspiration, he pricks his finger on the elusive quarry; a needle! Cantor, by contrast, stays indoors using pure reason to show that the mass of the haystack exceeds the mass of the hay in it. He deduces that there must be something else, that is, a needle, to account for the excess. Unlike Liouville, he remains cool and spotless.

Some mathematicians were troubled by a nonconstructive proof that relied upon the properties of infinite sets. Compared to Liouville's lengthy argument, Cantor's seemed too easy, almost like sleight-of-hand. The young Bertrand Russell (1872–1970) may not have been alone in his initial reaction to Cantor's ideas:

> I spent the time reading Georg Cantor, and copying out the gist of him into a notebook. At that time I falsely supposed all his arguments to be fallacious, but I nevertheless went through them all in the minutest detail. This stood me in good stead when later on I discovered that all the fallacies were mine [9].

Like Russell, mathematicians came to appreciate Cantor for the innovator he was. His 1874 paper ushered in a new era for analysis, where the ideas of set theory would be employed alongside the $\varepsilon - \delta$ arguments of the Weierstrassians.

Cantor's work had consequences, many of which were truly astonishing. For instance, it is easy to show that if the algebraic numbers and the transcendental numbers are *each* denumerable, then so is their union, the set of all real numbers. Because this is not so, Cantor knew that the transcendentals form a nondenumerable set and thus far outnumber their algebraic cousins. Eric Temple Bell put it this way: "The algebraic numbers are spotted over the plane like stars against a black sky; the dense blackness is the firmament of the transcendentals" [10]. This is a delightfully unexpected realization, for the plentiful numbers seem scarce, and the scarce ones seem plentiful. In a sense, Cantor showed that the transcendentals are the hay and not the needles.

A related but more far-reaching consequence was the distinction between "small" and "large" infinite sets. Cantor proved that a denumerable set, although infinite, was *insignificantly* infinite when compared to a nondenumerable counterpart. As his ideas took hold, mathematicians came to regard denumerable sets as so much jetsam, easily expendable when addressing questions of importance.

As we shall see, dichotomies between large and small sets would arise in other analytic settings. At the turn of the nineteenth century, René Baire found a "large/small" contrast in what he called a set's "category," and Henri Lebesgue found another in what he called its "measure." Although cardinality, category, and measure are distinct concepts, each provided a means of comparing sets that would prove valuable in mathematical analysis.

Cantor addressed other questions about infinite sets. One was, "Are there nondenumerable sets having greater cardinality than intervals?" This he answered in the affirmative. Another was, "Are there infinite sets of an intermediate cardinality between a denumerable sequence and a nondenumerable interval?" This he never succeeded in resolving. With Cantor's founding vision and continuing research, set theory took on a life of its own, quite apart from the concerns of analysis proper. But it all grew out of his 1874 paper.

Unlike many revolutionaries down through history, Georg Cantor lived to see his ideas embraced by the wider community. An early enthusiast was Russell, who described Cantor as "one of the greatest intellects of the nineteenth century" [11]. This is no small praise from a mathematician, philosopher, and eventual Nobel laureate.

Another of Cantor's admirers was the Italian prodigy Vito Volterra. His work, which beautifully combined Weierstrassian analysis and Cantorian set theory, is the subject of our next chapter.

CHAPTER 12

Volterra

Vito Volterra

Vito Volterra (1860–1940) flourished alongside a number of Italian mathematicians in the second half of the nineteenth century. Like his countrymen Giuseppe Peano (1858–1932), Eugenio Beltrami (1835–1900), and Ulisse Dini (1845–1918), he left his mark, contributing to applied areas like electrostatics and fluid dynamics, as well as to theoretical ones like mathematical analysis. It is of course the last of these that we consider here.

Although born on the Adriatic coast, Volterra was raised in Florence, the epicenter of the Italian Renaissance. He walked the same streets as had Michelangelo and attended schools named after Dante and Galileo. The fifteenth and sixteenth century Florentine atmosphere seems to have seeped into his bones, for Volterra loved art, literature, and music even as he loved science. He was a Renaissance Man, albeit three centuries removed.

Besides these pursuits, his political courage deserves to be celebrated. Witnessing the rise of Mussolini in the 1920s, Volterra took a public stand in opposition and signed a declaration against the regime. This act ultimately cost him his job but made him a hero for Italian intellectuals of the time. Upon his death in 1940, Italy had not yet shed its fascist scourge, but Volterra had fought the good fight in anticipation of a better future.

If he showed great courage late in life, he had shown great precocity early on. Young Volterra read college-level mathematics texts at age 11, impressed his teachers during adolescence, and somehow secured a position as a physics laboratory assistant at the University of Florence while still in high school. His academic career was spectacularly rapid, culminating with a doctorate in physics at the age of 22 [1].

In this chapter we discuss a pair of Volterra's early discoveries, both published in 1881, three years after his high school graduation. The first was another in the growing list of pathological counterexamples, one that turned up a previously unnoticed flaw in the Riemann integral. The second, almost paradoxically, was a theorem showing that pathology has its limits, for Volterra proved that no function can be continuous at each rational point and discontinuous at each irrational one. Such a function would simply be too pathological to exist. We shall examine the theorem in full, but we begin with a few words about the counterexample.

VOLTERRA'S PATHOLOGICAL FUNCTION

The second version of the fundamental theorem of calculus, which we saw in chapter 6, was stated by Cauchy as follows: "If F is differentiable and if its derivative F' is continuous, then $\int_a^b F'(x)dx = F(b) - F(a)$." Informally, this says that under the right conditions the integral of the derivative restores the original function. In the proof, Cauchy used the hypotheses that (a) F has a derivative and (b) this derivative is itself continuous. But were both necessary?

Statement (a) seems indispensable, for we could not hope to integrate a derivative if the derivative fails to exist. But the status of (b) is more suspect. Must we assume something as restrictive as the continuity of F' in order for the result to hold?

This is not a trivial issue. On the one hand, we saw in chapter 10 that the continuity of a derivative cannot be taken for granted, for the function

$$U(x) = \begin{cases} x^2 \sin(1/x) & \text{if } x \neq 0, \\ 0 & \text{if } x = 0, \end{cases}$$ has a discontinuous derivative. On the other hand, we do not need continuity to guarantee the existence of an integral, for it is easy to find discontinuous but integrable functions.

The question, then, was what condition, if any, we should impose upon F' to guarantee the truth of the fundamental theorem. Discoveries of the previous years gave mathematicians a perspective on the matter that Cauchy did not have, so it seemed worthwhile to revisit this important theorem.

In 1875, Gaston Darboux succeeded in weakening hypothesis (b). He proved that $\int_a^b F'(x)dx = F(b) - F(a)$ provided that (a) F is differentiable and (b') its derivative F' is Riemann integrable. Thus, we need not assume the continuity of F'; the mere existence of $\int_a^b F'(x)dx$ is sufficient for the fundamental theorem to hold.

This was progress of a sort, but there remained the issue of whether we need to assume *anything* about F' other than its existence. Perhaps derivatives are integrable by their very nature. If so, we could jettison both hypotheses (b) and (b') and build the fundamental theorem of calculus upon the assumption of (a) alone. That would be a less restrictive, and much more elegant, state of affairs.

It came down to this: How ill behaved can a derivative be? In an earlier chapter, we proved Darboux's theorem that a derivative, even if not continuous, must possess the intermediate value property. In that regard, derivatives seemed fairly "tame," and mathematicians might guess that such tameness would include integrability.

It was this misconception that the young Volterra refuted in his 1881 paper "*Sui principii del calcolo integrale*" [2]. There he provided an example of a function F that had a bounded derivative at all points but whose derivative was so discontinuous as to be nonintegrable. In other words, even though F was everywhere differentiable and its derivative F' was bounded, the integral $\int_a^b F'(x)dx$ did not exist. And, because the integral failed to exist, the equation $\int_a^b F'(x)dx = F(b) - F(a)$ could not be true. Volterra's example was striking not because the left-hand side of this equation was different from the right-hand side, but because the left-hand side was *meaningless*!

We shall not consider his function in detail, in part because it is complicated and in part because one chapter devoted to a pathological function

(Weierstrass's) may be enough. The interested reader will find a discussion of Volterra's work in [3].

One thing was clear: another unfortunate feature of the Riemann integral had been unearthed. Mathematicians would have loved nothing more than an uncluttered theorem to the effect that if F is differentiable with a bounded derivative F', then $\int_a^b F'(x)dx = F(b) - F(a)$. Volterra showed that, so far as Riemann's integral was concerned, this was not to be.

How could mathematicians respond to Volterra's strange example? One option was to accept the outcome and move on. When applying the fundamental theorem, we would simply impose an extra assumption about the derivative F'. This was the path of least resistance.

There was, however, an alternative. As we saw earlier, Riemann's integral provided no guarantee that $\lim_{k\to\infty} \int_a^b f_k(x)dx = \int_a^b \left[\lim_{k\to\infty} f_k(x)\right]dx$. Now Volterra had destroyed any hope for a simple fundamental theorem of calculus. As the nineteenth century neared its end, there was more reason than ever to suspect that the trouble lay in Riemann's definition and not in the intrinsic nature of analysis. A few daring souls, motivated in part by Volterra's pathological function, were about to forsake the Riemann integral in order to salvage the theorems above. Stay tuned.

HANKEL'S TAXONOMY

By the 1880s, mathematical analysis was awash in pathological counterexamples, each seemingly stranger than the last. Among those we have seen are:

(a) Dirichlet's function $\phi(x) = \begin{cases} c & \text{if } x \text{ is rational,} \\ d & \text{if } x \text{ is irrational,} \end{cases}$ which is every-

where discontinuous and not Riemann integrable.

(b) The extended ruler function R, which is continuous at each irrational and discontinuous at each rational but also is Riemann integrable with $\int_0^1 R(x)dx = 0$.

(c) Weierstrass's pathological function $f(x) = \sum_{k=0}^{\infty} b^k \cos(\pi a^k x)$, which is everywhere continuous and nowhere differentiable.

The situation suggested analytic chaos and cried out for order to be imposed upon so disorderly a mathematical scene.

One who tried to do just that was Hermann Hankel (1839–1873). He was an admirer of Riemann who believed that functions should be classified in a manner familiar to biologists or geologists. He proposed such a classification in 1870, a few years before his untimely death. With this taxonomy, he hoped to clarify the nature and limitations of mathematical analysis.

Hankel considered the family of all bounded functions defined on an interval [a, b] and distinguished them by means of their continuity/discontinuity properties. To see how he proceeded, we recall a familiar definition of Georg Cantor.

Definition: A set A of real numbers is *dense* if any open interval contains at least one member of A.

Elementary examples of dense sets are the rationals and the irrationals because any open interval holds infinitely many of both. The name is suggestive, for members of a dense set are so tightly packed that they are always nearby.

With this in mind, we are ready for Hankel's classification. In class 1 he placed those functions continuous at all points of [a, b]. These were well behaved in that they assumed maximum and minimum values, possessed the intermediate value property, and could be integrated. In Hankel's taxonomy, class 1 represented the top of the food chain.

His second class included functions continuous except at finitely many points of [a, b]. These were more problematic, but their irregularities, being finite in number, remained largely under control. One example is $S(x) = \begin{cases} \cos(1/x) & \text{if } x \neq 0, \\ 0 & \text{if } x = 0, \end{cases}$ defined on [−1, 1] because, as we saw in chapter 10, it has a single discontinuity at $x = 0$. Alternately, one could take a continuous function on an interval [a, b] and redefine it at, say, fifty points in order to introduce fifty discontinuities. Such a function would fall into Hankel's class 2.

Logically, there was but one class left: those functions possessing infinitely many points of discontinuity in [a, b]. These, of course, were the worst, but Hankel believed that they could be subdivided into the bad and the very bad:

Class 3A: Functions discontinuous at infinitely many points of [a, b] but still continuous on a dense set. These he called "pointwise discontinuous."

Class 3B: Everything else. These Hankel called "totally discontinuous."

We see that a pointwise discontinuous function in class 3A, in spite of its infinitude of discontinuities, must be continuous *somewhere* in any open interval. On the other hand, for a function in class 3B there must exist some open subinterval (c, d) within (a, b) where the function has no point of continuity at all. A totally discontinuous function thus features a solid subinterval featuring nothing but points of discontinuity.

Where do the three pathological functions cited above fit into Hankel's scheme? Dirichlet's function, being discontinuous everywhere, falls into class 3B as totally discontinuous. The ruler function is discontinuous at infinitely many points (the rationals) yet continuous on a dense set (the irrationals) and consequently belongs to class 3A as pointwise discontinuous. And Weierstrass's function, perhaps the weirdest of all, is paradoxically in class 1, for it is continuous everywhere.

Hankel found his classification important in the following sense: he knew that functions in class 1 and in class 2 are Riemann integrable, and the examples at his fingertips of pointwise discontinuous functions were integrable as well. By contrast, Dirichlet's totally discontinuous function was not. To him, the gap between classes 3A and 3B seemed to be the unbridgeable chasm. As Thomas Hawkins put it, "By making the distinction between pointwise and totally discontinuous functions, Hankel believed he had separated the functions amenable to mathematical analysis from those beyond its reaches" [4].

To demonstrate the value of all this, Hankel proved a spectacular theorem: a bounded function on $[a, b]$ was Riemann integrable if and only if it was no worse than pointwise discontinuous. That is, provided it fell into class 1, class 2, or class 3A, a bounded function could be integrated; those that occupied class 3B were not integrable and, by extension, analytically hopeless.

Hankel's theorem appeared to answer the major question we introduced earlier: "How discontinuous can an integrable function be?" The answer, according to him, was, "at worst pointwise discontinuous." His proof showed that, so long as a function was continuous on a dense set, all those discontinuities would not matter in terms of integrability. This was exactly the kind of simple result mathematicians had longed for.

Unfortunately, it was also incorrect.

With ideas this complicated, even great scholars can make mistakes, and Hankel made a doozy. To be fair, half of his theorem was true: if a function is Riemann integrable, it must indeed be continuous on a dense set. A totally discontinuous function, having a solid subinterval of points of discontinuity, cannot possess a Riemann integral. Again, one thinks of Dirichlet's function in this regard.

But Hankel's proof of the converse was flawed. In 1875, the British mathematician H. J. S. Smith (1826–1883) published an example of a pointwise discontinuous but non-integrable function which, he said, "deserves attention because it is opposed to a theory of discontinuous functions which has received the sanction of an eminent geometer, Dr. Hermann Hankel, whose recent death at an early age is such a great loss to mathematical science" [5]. Smith's example was nontrivial, requiring the construction of what we now call a nowhere dense set of positive measure. We refer those seeking details to Hawkins [6]. For now, we merely observe that the link between continuity and Riemann integrability remained unclear, and the question of how discontinuous an integrable function could be was still open. Pointwise discontinuity, whatever its value, did not provide the long-sought connection.

Nonetheless there had been progress of a sort. Riemann had extended the notion of integrability to include some highly discontinuous functions, and the true half of Hankel's theorem, along with Smith's counterexample, showed that the Riemann-integrable functions were properly embedded within the larger collection of functions that were continuous on a dense set.

We note in passing that the term "pointwise discontinuous" has sometimes been carelessly taken to mean "at worst pointwise discontinuous." That is, all functions in Hankel's classes 1, 2, or 3A were lumped under the single rubric of pointwise discontinuity, which led to the bizarre situation of placing the continuous functions (class 1) among the "pointwise discontinuous" ones. Because the common property of functions in these first three classes is that each is continuous on a dense set, we might suggest *densely continuous* as an umbrella term to include all functions in classes 1, 2, and 3A.

In any case, Hankel's taxonomy initially seemed to be a promising vehicle for carving apart the analytically accessible functions from the analytically intractable ones. As it turned out, however, many of those intractable functions could be handled quite nicely within the context of set theory and the Lebesgue integral. Nowadays, Hankel's distinctions have largely fallen by the wayside.

But in the late nineteenth century, pointwise discontinuity remained a topic of research capable of engaging the most talented mathematicians. One of these was the 21-year-old Vito Volterra.

THE LIMITS OF PATHOLOGY

The epidemic of pathological functions suggested that any behavior, no matter how bizarre, could be realized by an ingeniously constructed example from a suitably inventive mathematician.

Who, for instance, could envision the ruler function, continuous at each irrational point and discontinuous at each rational one? And why not suppose that somewhere, waiting to be discovered, lay an equally peculiar function continuous at each rational point and discontinuous at each irrational? One seemed no more outlandish than the other.

That continuity and discontinuity points can sometimes be interchanged is evident in the following examples. First define $H(x) = \begin{cases} x & \text{if } x \neq 0, \\ 1 & \text{if } x = 0. \end{cases}$ This is obviously continuous at all points but the origin, where it has its lone point of discontinuity.

As its counterpart, we introduce $K(x) = \begin{cases} x^2 & \text{if } x \text{ is rational}, \\ 0 & \text{if } x \text{ is irrational}. \end{cases}$ It is not difficult to see that K is discontinuous at any $a \neq 0$. For, if we let $\{x_k\}$ be a sequence of rationals converging to a and $\{y_k\}$ be a sequence of irrationals converging to a, then $\lim_{k \to \infty} K(x_k) = \lim_{k \to \infty}(x_k^2) = a^2$, whereas $\lim_{k \to \infty} K(y_k) = \lim_{k \to \infty} 0 = 0 \neq a^2$. Because these sequential limits differ, we know that $\lim_{x \to a} K(x)$ cannot exist and so K is discontinuous at $x = a$.

However, for any x, be it rational or irrational, we have $0 \leq K(x) \leq x^2$, and so a simple squeezing argument shows that $\lim_{x \to 0} K(x) = 0 = K(0)$. It follows that K is a function with a lone point of continuity: the origin. So, for H and K as defined here, the points of continuity and of discontinuity have been swapped.

In this regard, it will be useful to introduce the following.

Definition: For a function f, we let $C_f = \{x | f \text{ is continuous at } x\}$ and $D_f = \{x | f \text{ is discontinuous at } x\}$.

Our previous discussion can be neatly summarized by: $C_H = \{x | x \neq 0\} = D_K$ and $C_K = \{0\} = D_H$.

The issue of interchanging continuity and discontinuity points is an intriguing one. For any function f, is there a "complementary" function g

with $C_f = D_g$ and $C_g = D_f$? If so, how would one find it? If not, what would prevent it?

In his 1881 paper, "*Alcune osservasioni sulle funzioni punteggiate discontinue,*" Volterra addressed this matter. The result was a powerful theorem with a pair of first-rate corollaries [7].

Theorem: There cannot exist two pointwise discontinuous functions on the interval (a, b) for which the continuity points of one are the discontinuity points of the other, and vice versa.

Proof: He proceeded by contradiction, assuming at the outset that f and ϕ are pointwise discontinuous on (a, b) such that $C_f = D_\phi$ and $D_f = C_\phi$. In other words, C_f and C_ϕ partition (a, b) into nonempty, disjoint, dense subsets.

His proof rested upon a nested sequence of subintervals. Because f is pointwise discontinuous, it must have a point of continuity x_0 somewhere in (a, b). For $\varepsilon = 1/2$, continuity guarantees that there exists a $\delta > 0$ so that $(x_0 - \delta, x_0 + \delta)$ is a subset of (a, b) and, if $0 < |x - x_0| < \delta$, then $|f(x) - f(x_0)| < 1/2$. We now choose $a_1 < b_1$ so that $[a_1, b_1]$ is a closed subinterval of the open set $(x_0 - \delta, x_0 + \delta)$, as depicted in figure 12.1.

For any two points x and y in $[a_1, b_1]$, we apply the triangle inequality to see that

$$|f(x) - f(y)| \leq |f(x) - f(x_0)| + |f(x_0) - f(y)| < 1/2 + 1/2 = 1. \qquad (1)$$

This means that f does not oscillate more than 1 unit on the closed interval $[a_1, b_1]$.

But (a_1, b_1) is an open subinterval of (a, b) and ϕ is pointwise discontinuous as well. Thus there is a point of continuity of ϕ, say x_1, within (a_1, b_1). Repeating the previous argument for ϕ, we find points $a_1' < b_1'$ such that the closed interval $[a_1', b_1']$ is a subset of (a_1, b_1) and $|\phi(x) - \phi(y)| < 1$ for any x and y in $[a_1', b_1']$. See figure 12.2.

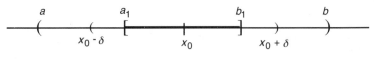

Figure 12.1

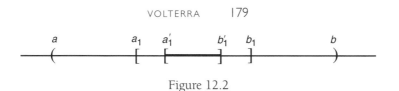

Figure 12.2

Combining this conclusion with that of (1) above, we have found a closed subinterval $[a'_1, b'_1]$ so that, for all x and y within it,

$$|f(x) - f(y)| < 1 \text{ and } |\phi(x) - \phi(y)| < 1.$$

Volterra then exploited pointwise discontinuity to repeat the argument with $\varepsilon = 1/4$. Considering first f and then ϕ, he found a closed interval $[a'_2, b'_2]$ lying within the open interval (a'_1, b'_1)—and thus inside $[a'_1, b'_1]$—such that $|f(x) - f(y)| < 1/2$ and $|\phi(x) - \phi(y)| < 1/2$ for any points x and y in $[a'_2, b'_2]$.

He continued with $\varepsilon = 1/8$, $1/16$, and generally $1/2^k$, thereby generating closed intervals $[a'_1, b'_1] \supset [a'_2, b'_2] \supset [a'_3, b'_3] \supset \cdots$ such that

$$|f(x) - f(y)| < 1/2^{k-1} \text{ and } |\phi(x) - \phi(y)| < 1/2^{k-1}$$

$$\text{for any } x \text{ and } y \text{ in } [a'_k, b'_k]. \tag{2}$$

A contradiction was at hand. By the completeness property, there must be a point c common to all of the nested intervals $[a'_k, b'_k]$. Because c lies in $[a'_1, b'_1]$, it is indeed in our original interval (a, b).

We next claim that f is continuous at c. This follows easily, for Volterra had controlled the oscillation of f as he constructed his descending intervals. To be thoroughly Weierstrassian about it, we could take any $\varepsilon > 0$ and choose a whole number k so that $1/2^{k-1} < \varepsilon$. We know that c is a point of $[a'_{k+1}, b'_{k+1}]$, which in turn lies within the open interval (a'_k, b'_k) so we can find a $\delta > 0$ with $(c - \delta, c + \delta) \subset (a'_k, b'_k) \subset [a'_k, b'_k]$. Consequently, for any x with $0 < |x - c| < \delta$, then by (2), we have $|f(x) - f(c)| < 1/2^{k-1} < \varepsilon$. This proves that $\lim_{x \to c} f(x) = f(c)$, and so f is continuous at c as claimed.

Because the same argument, word for word, can be applied to ϕ, it too is continuous at c. In this way, we have reached our contradiction, for c belongs to both C_f and C_ϕ, violating the hypotheses that the continuity points of one are the discontinuity points of the other. There is no alternative but to conclude that two such pointwise discontinuous functions cannot exist.   Q.E.D.

Before proceeding, we make a pair of observations. The first is that Volterra was vague about insisting that the intervals $[a'_k, b'_k]$ be *closed*. This is an omission easily repaired, as we have done. Second, in the example above where the continuity points of H are the discontinuity points of K and vice versa, we note that K is totally discontinuous (Hankel's class 3B) rather than pointwise discontinuous (Hankel's class 3A). Consequently— lest anyone lose sleep on this account—that example in no way contra- dicts Volterra's result.

He followed his theorem with two important corollaries. The first, which settled a major question of analysis, was stated as follows:

> Because we have a function continuous at each irrational point and discontinuous at each rational, it will be impossible to find a function that is discontinuous at each irrational point and contin- uous at each rational. [8]

To flesh out his argument, we imagine a function G for which C_G is the (dense) set of rationals. Then G is pointwise discontinuous. But we have previously encountered the extended ruler function R which is pointwise discontinuous as well, with C_R being the set of irrationals. The continuity points of G would then be the discontinuity points of R, in con- tradiction to Volterra's theorem. Consequently, it is impossible for both functions to exist. Because the ruler function most certainly *does* exist, we are forced to conclude that the function G does not. Volterra's theorem demonstrated, in the parlance of a Western movie, that "this town is not big enough for both of them." A function continuous only on the rationals is a logical impossibility.

Pathology, then, has its limits. No matter how clever the mathemati- cian, certain functions remain beyond the pale, a fact Volterra demonstrated with this clever argument. But he had one more corollary up his sleeve, that there can be no continuous function taking rationals to irrationals and vice versa [9].

Corollary: There does not exist a continuous function g defined on the real numbers such that $g(x)$ is rational when x is irrational and $g(x)$ is irrational when x is rational.

Proof: Again, for the sake of contradiction, Volterra assumed such a function g exists. We then define G by $G(x) = R(g(x))$, where R is the extended ruler function from above, and make two claims about G:

Claim 1: If x is rational, G is continuous at x.

This is evident because, if x is rational, $g(x)$ is irrational, so R is continuous at $g(x)$. But g is assumed to be continuous everywhere, so the composite function G will be continuous at x.

Claim 2: If y is irrational, then G is discontinuous at y.

This is easily verified by choosing a sequence $\{x_k\}$ of rationals converging to y. Then

$$\lim_{k\to\infty} G(x_k) = \lim_{k\to\infty} R(g(x_k)) = \lim_{k\to\infty} 0 = 0,$$

because g carries each rational x_k to an irrational $g(x_k)$, and the ruler function is zero at irrational points. On the other hand, $G(y) = R(g(y)) \neq 0$ because $g(y)$ is rational. In short, $\lim_{k\to\infty} G(x_k) \neq G(y)$, and so G is discontinuous at y.

Taken together, these claims show that G is continuous upon the rationals and discontinuous upon the irrationals—a situation that Volterra had just proved to be impossible! It follows that a function like g cannot exist. There is no continuous transformation that carries rationals to irrationals and vice versa.   Q.E.D.

Among other things, these results remind us that the rationals and irrationals, although both dense sets of real numbers, are intrinsically noninterchangeable. As we saw, Cantor had highlighted the fact that the rationals are denumerable and the irrationals are not, but mathematicians would find other, more subtle distinctions between these systems. One of these was the notion of a set's "category," a concept due to Volterra's gifted student René Baire, who is the subject of our next chapter.

With this, we leave the 21-year-old Vito Volterra. A long and distinguished career lay ahead of him, one that would see continued mathematical success, international recognition, and even an honorary knighthood from Britain's King George V.

Looking back from later in his life, Volterra characterized the 1800s as "the century of the theory of functions" [10]. Starting with Euler's initial ideas, the concept of function had assumed a central role in the work of Cauchy, Riemann, and Weierstrass and then been passed to the generation of Cantor, Hankel, and Volterra himself. Functions had come to dominate analysis, and their unexpected possibilities surprised mathematicians time and again. As we have seen, Volterra deserves a place in this tale for two different but fascinating discoveries from 1881.

For such a young man, it had been quite a year.

CHAPTER 13

Baire

René Baire

In his doctoral thesis of 1899, René Baire (1874–1932) assessed the importance of set theory to mathematical analysis:

> One can even say, in a general manner, that . . . any problem relative to the theory of functions leads to certain questions relative to the theory of sets and, insofar as these latter questions are or can be addressed, it is possible to resolve, more or less completely, the given problem [1].

As we shall see, Baire not only advocated this position but did a splendid job of practicing it.

Unfortunately, his mathematical triumphs were confined to the brief periods when he was both physically and mentally sound. An introverted person of "delicate" health, Baire entered university in 1892, and his obvious talents took him to Italy to study with Volterra [2]. After completing his dissertation, *Sur les fonctions de variables réelles*, Baire taught at the Universities of Montpellier (1902) and Dijon (1905). During this time, despite the occasional setback, Baire seemed able to cope.

But then a series of ailments destroyed his fragile constitution. He endured everything from restrictions of the esophagus to severe attacks of agoraphobia. By 1909 his teaching had deteriorated beyond repair, and in 1914 he was given a leave of absence from Dijon. Baire would never return to serious research.

Instead, he spent his remaining years fighting physical and mental demons while burdened with sometimes crushing poverty. A colleague described him as "the type of man of genius who pays for that genius with a continual suffering due to an always unsteady constitution" [3]. In all, René Baire had only a dozen good years to devote to mathematics.

In this chapter, we shall look back to his dissertation and the first appearance of what is now known as the Baire category theorem. We begin, as did Baire, with the concept of a nowhere-dense set.

NOWHERE-DENSE SETS

As noted earlier, a set of real numbers is *dense* if every open interval contains at least one member of the set. In modern notation, D is dense if, for any open interval (α, β), we have $(\alpha, \beta) \cap D \neq \varnothing$.

A set fails to be dense if there is an open interval containing no points of the set. For instance, we let E be the set of all *positive* rational numbers. This is not dense in the real line because the open interval $(-2, 0)$ is free of points of E. However, E exhibits a "denseness" over part of its reach, for members of E are present in any open interval (α, β) where $0 < \alpha < \beta$.

In order to move beyond examples like this, that is, those that are dense in some regions but not in others, we introduce a new idea.

Definition: A set P of real numbers is *nowhere dense* if every open interval (α, β) contains an open subinterval $(a, b) \subseteq (\alpha, \beta)$ such that $(a, b) \cap P = \varnothing$.

This means that, even though points of P might be found in a given interval (α, β), there is an entire subinterval within it that is free of such points (see figure 13.1). Nowhere-dense sets are thus regarded as being sparse or, to use the descriptive term of Hermann Hankel, "scattered" [4].

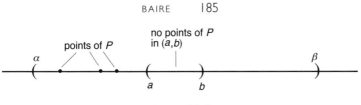

Figure 13.1

We note that "nowhere dense" is not the logical negation of "dense." The nondense set E above, for instance, is *not* nowhere dense because the open interval $(3, 4)$ contains no subinterval free of positive rationals. We thus would do well to provide a few examples of sets that *are* nowhere dense.

1. The set consisting of a single point $\{c\}$ is nowhere dense.
This is obvious, for if (α, β) is an open interval not containing c, then $(\alpha, \beta) \subseteq (\alpha, \beta)$ and $(\alpha, \beta) \cap \{c\} = \varnothing$. On the other hand, if (α, β) is an open interval containing c, then $(c, \beta) \subseteq (\alpha, \beta)$ and $(c, \beta) \cap \{c\} = \varnothing$.

2. The set $S = \left\{ \dfrac{1}{k} \middle| k \text{ is a whole number} \right\} = \left\{ 1, \dfrac{1}{2}, \dfrac{1}{3}, \dfrac{1}{4}, \ldots \right\}$ is nowhere dense.

This too is easy to see, for the gaps between reciprocals of two consecutive integers will furnish subintervals free of points of S. Even if a given open interval (α, β) contains 0—the point towards which these reciprocals are accumulating—we can choose a whole number N so that $\dfrac{1}{N} \in (\alpha, \beta)$ and take the open subinterval $\left(\dfrac{1}{N+1}, \dfrac{1}{N} \right) \subseteq (\alpha, \beta)$ with $\left(\dfrac{1}{N+1}, \dfrac{1}{N} \right) \cap S = \varnothing$, as shown in figure 13.2.

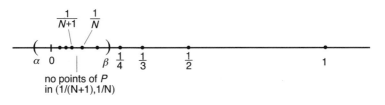

Figure 13.2

3. The set $T = \left\{ \dfrac{1}{r} + \dfrac{1}{k} \middle| r \text{ and } k \text{ are whole numbers} \right\}$ is nowhere dense.

To conjure up a mental picture of this set, fix r and let k run through the positive integers. This generates points $\dfrac{1}{r} + 1, \dfrac{1}{r} + \dfrac{1}{2}, \dfrac{1}{r} + \dfrac{1}{3}, \dfrac{1}{r} + \dfrac{1}{4}, \ldots$, which cluster around $\dfrac{1}{r}$ in the same way that the points of the previous example clustered around 0. Because r is arbitrary, *every* reciprocal $\dfrac{1}{r}$ is such a cluster point, giving T quite a complicated structure. Nonetheless, the gaps among the points $\dfrac{1}{r} + \dfrac{1}{k}$ are such as to make T nowhere dense (we omit the details).

Before seeing what Baire made of this, we prove two simple lemmas that will come in handy.

Lemma 1: Subsets of nowhere-dense sets are nowhere dense. That is, if P is a nowhere-dense set and $U \subseteq P$, then U is nowhere dense.

Proof: Given an open interval (α, β), we know there exists an open subinterval $(a, b) \subseteq (\alpha, \beta)$ with $(a, b) \cap P = \emptyset$. Because U is a subset of P, it is clear that $(a, b) \cap U = \emptyset$, and so U is nowhere dense as well. Q.E.D.

Lemma 2: The union of two nowhere-dense sets is nowhere dense.

Proof: Let P_1 and P_2 be nowhere dense. To show that $P_1 \cup P_2$ is also nowhere dense, we begin with an open interval (α, β). Because P_1 is nowhere dense, there exists an open subinterval $(a, b) \subseteq (\alpha, \beta)$ with $(a, b) \cap P_1 = \emptyset$. But (a, b) is itself an open interval and P_2 is nowhere dense, so there is an open subinterval $(c, d) \subseteq (a, b) \subseteq (\alpha, \beta)$ with $(c, d) \cap P_2 = \emptyset$. Clearly, (c, d) is an open subinterval of (α, β) containing no points of P_1 or P_2. Thus, $(c, d) \cap (P_1 \cup P_2) = \emptyset$, so $P_1 \cup P_2$ is nowhere dense. Q.E.D.

As this second lemma shows, we can amalgamate two—or for that matter any finite number—of nowhere-dense sets and still find ourselves with a nowhere-dense outcome. Even the union of a million such sets would remain, in Hankel's terminology, scattered.

But what if we assemble an *infinitude* of nowhere-dense sets? What sort of structure might such a union have? And what use might this be to mathematical analysis? These are matters that Baire addressed with his characteristic ingenuity.

THE BAIRE CATEGORY THEOREM

In his thesis, Baire wrote of a set F with the property that

there exists a denumerable infinity of sets $P_1, P_2, P_3, P_4, \ldots$, each nowhere dense, such that every point [of F] belongs to at least one of the sets $P_1, P_2, P_3, P_4, \ldots$. I will say a set of this nature is of the *first category*. [5]

In other words, F is a set of the first category if $F = P_1 \cup P_2 \cup P_3 \cup \cdots \cup P_k \cup \cdots$, where each P_k is nowhere dense.

Many later mathematicians have been critical of Baire not for his ideas but for his terminology. The completely nondescriptive "first category" is about as colorless a term as there is and conjures up no image whatever in the mind's eye. Such critics must have been further dismayed when they read on: "Any set which does not possess this property [first category] will be said to be of the *second category*."

It is clear that a denumerable set is of the first category. Such a set $\{a_1, a_2, a_3, a_4, \ldots\}$ can be written as the union of one-point sets

$$\{a_1\} \cup \{a_2\} \cup \cdots \cup \{a_k\} \cup \cdots,$$

where, as we saw, each one-point set is nowhere dense. In particular, this means that the (denumerable) set of algebraic numbers is of the first category, as is its (denumerable) subset, the rationals. But the rationals form a dense set. So, whereas finite unions of nowhere-dense sets must remain nowhere dense, denumerable unions of such sets can grow sufficiently large to be everywhere dense. As Baire put it, a first category set "can evidently be of a different nature than the individual sets P_k" [6]. If we agree that nowhere-dense sets are "small," are we ready to conclude that first category sets are, for want of a better word, "large"?

Before seeing what Baire had to say about this, we need a few more lemmas.

Lemma 3: Any subset of a first category set is itself of the first category.

Proof: Let $F = P_1 \cup P_2 \cup P_3 \cup \cdots \cup P_k \cup \cdots$ be of the first category, where each P_k is nowhere dense, and let $G \subseteq F$. Elementary set theory shows that

$$G = G \cap F = (G \cap P_1) \cup (G \cap P_2) \cup \cdots \cup (G \cap P_k) \cup \cdots,$$

where each $G \cap P_k$ is a subset of P_k and so is nowhere dense by lemma 1. Because G is then a denumerable union of nowhere-dense sets, it is of the first category.   Q.E.D.

We remark that lemma 3 implies that if S is a set of the second category and $S \subseteq T$, then T must also be of the second category. Just as shrinking a first category set yields another of that category, so too does enlarging a second category set result in another second category set.

Lemma 4: The union of two first category sets is first category.

Proof: Let F and H be of the first category. Then $F = P_1 \cup P_2 \cup P_3 \cup \cdots \cup P_k \cup \cdots$, where each P_k is nowhere dense, and $H = R_1 \cup R_2 \cup \cdots \cup R_k \cup \cdots$, where each R_k is nowhere dense. We shuffle these sets together to write

$$F \cup H = (P_1 \cup R_1) \cup (P_2 \cup R_2) \cup \cdots \cup (P_k \cup R_k) \cup \cdots,$$

and each set $P_k \cup R_k$ is nowhere dense by lemma 2. Thus, $F \cup H$ is the denumerable union of nowhere-dense sets and so is of the first category.   Q.E.D.

Lemma 4 rests upon the fact that the union of two denumerable collections is denumerable, and we can extend this to three or four or any finite number of such collections. Better yet, the *denumerable* union of denumerable collections is denumerable, so we have the following lemma.

Lemma 5: If $F_1, F_2, \ldots, F_k, \ldots$ is a denumerable collection of sets of the first category, then their union $F_1 \cup F_2 \cup \cdots \cup F_k \cup \cdots$ is of the first category as well.

As noted, the dense set of rationals is of the first category, suggesting that sets of this type may be "large." But appearances are deceptive. In 1899 Baire proved that a first category set must be, in a fundamental sense, "small." To be precise, such a set is never sufficient to exhaust an open interval. It is this result that now carries his name.

Theorem (Baire category theorem): If $F = P_1 \cup P_2 \cup P_3 \cup \cdots \cup P_k \cup \cdots$, where each P_k is a nowhere-dense set, and if (α, β) is an open interval, then there exists a point in (α, β) that is not in F.

Proof: We begin with (α, β) and consider the nowhere-dense set P_1. By definition, there is an open subinterval of (α, β) containing no points of P_1. By shrinking this subinterval if necessary, we can find $a_1 < b_1$ such that the closed subinterval $[a_1, b_1] \subseteq (\alpha, \beta)$ and $[a_1, b_1] \cap P_1 = \emptyset$. (We remark that Baire, like Cantor and Volterra before him, did not emphasize the need for *closed* subintervals.)

But (a_1, b_1) is itself an open interval and P_2 is nowhere dense, so in analogous fashion we have $a_2 < b_2$ with $[a_2, b_2] \subseteq (a_1, b_1) \subseteq [a_1, b_1] \subseteq (\alpha, \beta)$ and $[a_2, b_2] \cap P_2 = \emptyset$. Continuing in this way, we construct a descending sequence of closed intervals

$$[a_1, b_1] \supseteq [a_2, b_2] \supseteq \cdots \supseteq [a_k, b_k] \supseteq \cdots,$$

where $[a_k, b_k] \cap P_k = \emptyset$ for each $k \geq 1$.

By the nested interval version of the completeness property, there is at least one point c common to all of these intervals. To complete the proof, we need only show that c is a point of the open interval (α, β) not belonging to F.

First, because c is in all the closed intervals, $c \in [a_1, b_1] \subseteq (\alpha, \beta)$, and so c indeed lies within (α, β).

Second, for each $k \geq 1$, we know that c is in $[a_k, b_k]$ and that $[a_k, b_k]$ has no points in common with P_k. The point c, belonging to *none* of the P_k, cannot belong to their union, F.

We have thus found a point of (α, β) not contained in the first category set F. In short, a first category set cannot exhaust an open interval.   Q.E.D.

Je commence par démontrer la proposition suivante : Si P est un ensemble de première catégorie, il existe, dans toute portion $\alpha\beta$ du segment sur lequel il est défini, au moins un point (et par suite une infinité) n'appartenant pas à P. En effet, d'après les hypothèses, on peut déterminer dans $\alpha\beta$ un intervalle fini $\alpha_1\beta_1$ ne contenant aucun point de P_1; dans $\alpha_1\beta_1$, un intervalle $\alpha_2\beta_2$ ne contenant aucun point de P_2, etc....; dans $\alpha_{n-1}\beta_{n-1}$, un intervalle $\alpha_n\beta_n$ ne contenant aucun point des n premiers ensembles $P_1, P_2,...$ P_n; il existe au moins un point M compris à l'intérieur de tous les segments $\alpha_n\beta_n$; ce point M ne fait partie d'aucun ensemble P_n et par suite ne fait pas partie de P.

The Baire category theorem (1899)

This is the original proof of the Baire category theorem. His elegant argument used the completeness property and did so in a manner reminiscent of the result we have seen from his mentor Volterra. Baire continued:

It follows immediately that any interval is a set of the second category; for we have just proved that one cannot obtain all points of a continuous interval by means of a denumerable infinity of nowhere dense sets [7].

From this we can deduce that the set of all real numbers is of the second category, for the reals contain within them the second category set $(0, 1)$. And this means that the set of irrationals is of the second category, for otherwise, both the rationals and the irrationals would be of the first category, as would be their union by Lemma 3. But their union is all the real numbers, a second category set.

At this point, Baire contrasted sets of the first and second categories:

One sees the profound difference that exists between sets of the two categories; this difference does not reside in their denumerability nor in their condensation within an interval, for a set of the first category can have the cardinality of the continuum and can be dense; but it is in some sense a combination of the two preceding notions [8].

From what is now called the *topological* viewpoint, the Baire category theorem shows that first category sets are in a sense negligible. Some authors who object to Baire's colorless terminology use *meager* as a more suggestive alternative for "first category." Whatever their names, Baire's dichotomy would have important consequences for mathematical analysis, as the next section illustrates.

SOME APPLICATIONS

A hallmark of mathematical progress is the fruitful generalization, one that gathers seemingly unrelated matters under a single umbrella. Such a generalization is both more efficient and more elegant than what came before. The Baire category theorem is one of these, as is clear if we return to Cantor's nondenumerability result from chapter 11.

Cantor's Theorem Revisited: If $\{x_k\}$ is a sequence of distinct real numbers, then any open interval (α, β) contains a point not included among the $\{x_k\}$.

Proof: The collection $\{x_1, x_2, x_3, \ldots, x_k, \ldots\}$, considered as a set of points, is denumerable and thus of the first category. Because Baire showed that a first category set cannot exhaust an open interval, (α, β) must contain a point other than the $\{x_k\}$.   Q.E.D.

That was certainly easy.

But there is more. Volterra's major result from chapter 12 is also a consequence of Baire's work. To see this, we need some background, including an immediate corollary of the category theorem.

Corollary: The complement of a first category set is dense.

Proof: (Recall that the complement of a set of real numbers A, often denoted by A^c, is the set of real numbers not belonging to A.) Let F be of the first category and consider any open interval (α, β). Baire proved that not every point in (α, β) belongs to F, so $(\alpha, \beta) \cap F^c \neq \emptyset$, and this is precisely what it required to show that the complement of F is dense.   Q.E.D.

We next wish to characterize pointwise discontinuous functions in terms of category, a quest that had led Baire to investigate category in the first place. In what follows, we join Baire in adopting the "inclusive" meaning of pointwise discontinuity, that is, continuity on a dense set. But our discussion differs from his original in that he employed the function's oscillation, whereas we reach the same end by means of sequences [9].

Beginning with a function f and a whole number k, we define the set

$$P_k \equiv \left\{ x \,|\, \text{there is a sequence } a_j \to x \text{ with} |f(a_j) - f(x)| \geq \frac{1}{k} \text{ for all } j \geq 1 \right\}.$$

(1)

A real number x thus belongs to P_k if we can approach x sequentially by means of $\{a_j\}$ in such a way that the functional values $f(a_j)$ and $f(x)$ are all separated by a gap of at least $1/k$. As an example, we again consider the function

$$S(x) = \begin{cases} \cos(1/x) & \text{if } x \neq 0, \\ 0 & \text{if } x = 0 \end{cases}$$

from chapter 10 and claim that 0 belongs to the set P_2. To verify this, we introduce the sequence $\left\{ \dfrac{1}{2\pi j} \right\}$. Clearly $\displaystyle \lim_{j \to \infty} \frac{1}{2\pi j} = 0$, and for each $j \geq 1$ we have $\left| S\left(\dfrac{1}{2\pi j} \right) - S(0) \right| = |\cos(2\pi j) - 0| = 1 \geq \dfrac{1}{2}$. By the definition in (1), we see that $0 \in P_2$.

We are now ready to prove Baire's characterization of pointwise discontinuity in terms of the "smallness" of D_f.

Theorem: f is (at worst) pointwise discontinuous if and only if D_f is a set of the first category.

There are, of course, two implications to be proved. We begin with the more intricate necessary condition.

Necessity: If f is (at worst) pointwise discontinuous, then D_f is of the first category.

Proof: Our first object is to show that each P_k as defined above is nowhere dense. We thus fix a whole number $k \geq 1$ and an open interval (α, β). By pointwise discontinuity, f is continuous at some point—call it x_0—within

(α, β). This means that $\lim_{x \to x_0} f(x) = f(x_0)$, and so, for $\varepsilon = \dfrac{1}{3k}$, there exists a $\delta > 0$ such that the open interval $(x_0 - \delta, x_0 + \delta)$ is a subset of (α, β) and

$$\text{if } |x - x_0| < \delta, \text{ then } |f(x) - f(x_0)| < \frac{1}{3k}. \tag{2}$$

We assert that $(x_0 - \delta, x_0 + \delta) \cap P_k = \varnothing$. To prove this, suppose the opposite. Then there is some point z belonging to $(x_0 - \delta, x_0 + \delta) \cap P_k$. By the nature of P_k there must be a sequence $a_j \to z$ with $|f(a_j) - f(z)| \geq \dfrac{1}{k}$ for all $j \geq 1$. Because the sequence $\{a_j\}$ converges to $z \in (x_0 - \delta, x_0 + \delta)$, there exists a subscript N so that $a_N \in (x_0 - \delta, x_0 + \delta)$. With some help from the triangle inequality, we conclude that

$$\frac{1}{k} \leq |f(a_N) - f(z)| = |f(a_N) - f(x_0) + f(x_0) - f(z)|$$

$$\leq |f(a_N) - f(x_0)| + |f(x_0) - f(z)| < \frac{1}{3k} + \frac{1}{3k} = \frac{2}{3k},$$

where the last step follows from (2) and the fact that both $|a_N - x_0| < \delta$ and $|z - x_0| < \delta$. This chain of inequalities leaves us with the contradiction that $\dfrac{1}{k} < \dfrac{2}{3k}$. Something is amiss.

The trouble arose from the assumption that $(x_0 - \delta, x_0 + \delta) \cap P_k$ is nonempty. We conclude instead that $(x_0 - \delta, x_0 + \delta)$ is a subinterval of (α, β) that contains no points of P_k. By definition P_k is nowhere dense for each k, and this in turn means that $P_1 \cup P_2 \cup \cdots \cup P_k \cup \cdots$ is a set of the first category.

We are nearly done. We need only apply the notion of continuity—or, more precisely, of discontinuity—to see that

$$D_f \subseteq P_1 \cup P_2 \cup \cdots \cup P_k \cup \cdots. \tag{3}$$

Expression (3) follows because if $x \in D_f$ is any point of discontinuity of f, then there exists an $\varepsilon > 0$ so that, for any $\delta > 0$, we can find a point z with $0 < |z - x| < \delta$ yet $|f(z) - f(x)| \geq \varepsilon$. We then choose a whole number k with $\dfrac{1}{k} < \varepsilon$ and let δ equal, in turn, $1, \dfrac{1}{2}, \dfrac{1}{3}, \ldots$ to generate points $a_1, a_2, a_3, \ldots, a_j, \ldots$ with $0 < |a_j - x| < \dfrac{1}{j}$ but $|f(a_j) - f(x)|$

$\geq \varepsilon > \dfrac{1}{k}$. The sequence $\{a_j\}$ converges to x, yet for all $j \geq 1$, we have

$|f(a_j) - f(x)| > \dfrac{1}{k}$. By the definition in (1), the discontinuity point x belongs to the nowhere-dense set P_k and so, indeed, $D_f \subseteq P_1 \cup P_2 \cup \cdots \cup P_k \cup \cdots$.

We wrap up this half of the proof by noting that D_f, a subset of the first category set $P_1 \cup P_2 \cup \cdots \cup P_k \cup \cdots$, is itself first category by lemma 3. Therefore, if f is pointwise discontinuous, then D_f is a set of the first category.

Sufficiency: If D_f is of the first category, then f is (at worst) pointwise discontinuous.

Proof: This is an immediate consequence of the corollary to the Baire category theorem that we introduced earlier. Because D_f is of the first category, its complement is dense. In other words, $D_f^c = C_f = \{x \mid f$ is continuous at $x\}$ is dense, which is precisely what is required for f to be at worst pointwise discontinuous.   Q.E.D.

Thus the pointwise discontinuous functions are those whose assembled discontinuities remain "small" in the sense of being of the first category. This characterization reduced Hankel's thirty-year-old notion of pointwise discontinuity to a simple condition on the set D_f. Besides having its own intrinsic value, it allowed Baire to give an elegant proof of Volterra's theorem from the previous chapter [10].

Volterra's Theorem Revisited: There do not exist two pointwise discontinuous functions on the interval (a, b) for which the continuity points of one are the discontinuity points of the other, and vice versa.

Proof: Suppose for the sake of argument that f and ϕ were two such functions. The previous theorem shows that both D_f and D_ϕ are of the first category and so too is $D_f \cup D_\phi$ by lemma 4. By the Baire category theorem, the complement of this union is dense. But the complement in question is the set of points at which neither function is discontinuous, that is, the set of their common points of continuity. We have reached a contradiction, for f and ϕ share not just a single point of continuity but a *dense* set of them.   Q.E.D.

And, with little additional effort, Baire provided the following dramatic extension [11].

Theorem: If $f_1, f_2, \ldots, f_k, \ldots$ is a sequence of (at worst) pointwise discontinuous functions defined on a common interval, then there is a point—indeed, a dense set of points—at which all of these are simultaneously continuous.

Proof: As in the preceding proof, we consider D_{f_k}, the set of discontinuity points of the function f_k. By pointwise discontinuity, each of these is of the first category, and so their union $D_{f_1} \cup D_{f_2} \cup \cdots \cup D_{f_k} \cup \cdots$ is of the first category by Lemma 5. Again, the complement of this union is dense, but this complement is $C_{f_1} \cap C_{f_2} \cap \cdots \cap C_{f_k} \cap \cdots$ the points where all the functions are continuous at once.   Q.E.D.

This theorem shows that even though pointwise discontinuous functions can have infinitely many discontinuities, and even though we assemble a denumerable infinitude of such functions, enough continuity remains to guarantee that they share a dense set of points where all are continuous. This represents a perfect fusion of set theory and analysis, blended together under the watchful eye of René Baire.

Before leaving this section, we mention a last consequence Baire drew from his great theorem, one that led him to another lasting innovation [12].

Theorem: The uniform limit of pointwise discontinuous functions is pointwise discontinuous.

Here he began with a sequence $f_1, f_2, \ldots, f_k, \ldots$ of pointwise discontinuous functions defined on a common interval and assumed they converged uniformly to a function f. As we have seen, uniform convergence as described by Weierstrass was sufficiently strong to transfer certain properties from individual functions to their limit. Baire established that "pointwise discontinuity" was one such property.

Although omitting details, we give a sense of his argument. Under uniform convergence, Baire showed that any common point of continuity of the individual functions f_k must be a point of continuity of the limit function f. To put this in set-theoretic notation, he proved

$$C_{f_1} \cap C_{f_2} \cap \cdots \cap C_{f_k} \cap \cdots \subseteq C_f.$$

As we just saw, Baire knew that this denumerable intersection was dense, and so C_f must be dense as well. Then the uniform limit f, being continuous on a dense set, was pointwise discontinuous as claimed.

The fact that *uniform* limits of pointwise discontinuous functions must be pointwise discontinuous led Baire to wonder what, if anything, could be said about nonuniform limits. His reflections produced a new taxonomy of functions, much more sophisticated than Hankel's from a quarter-century earlier. We end the chapter with a discussion of these ideas.

THE BAIRE CLASSIFICATION OF FUNCTIONS

In the hope of categorizing functions into logically meaningful classes, Baire, like Hankel, took the continuous ones as his starting point. "I choose to say that the continuous functions constitute class 0," he wrote, in the process solidifying his reputation for colorless terminology [13].

Suppose we have a sequence of continuous, that is, class 0, functions $\{f_k\}$, and let $f(x) = \lim_{k \to \infty} f_k(x)$ be their pointwise limit. As we saw, f may or may not be continuous. In the latter case, the limit function has escaped from class 0, so Baire was ready with a new class. "Those discontinuous functions that are limits of continuous functions," he wrote, "form class 1." As an example, we recall from chapter 9 that each function $f_k(x) = (\sin x)^k$ is continuous on $[0, \pi]$, but $f(x) = \lim_{k \to \infty} f_k(x)$ is discontinuous at $\pi/2$. So, f belongs to class 1.

Baire proved something far more interesting: that functions in class 1 are at worst pointwise discontinuous [14]. That is, when we take a limit of continuous functions, the outcome need not be continuous everywhere but must at least be continuous on a dense set. Taking limits of continuous functions, then, cannot obliterate all vestiges of continuity. On the contrary, such limits retain a "respectable" amount of continuity from the originals. For those seeking a permanence in analysis, there is *some* comfort in that conclusion.

One consequence is the following.

Theorem: If f is differentiable, then its derivative f' must be continuous on a dense set.

Proof: For each $k \geq 1$, we define a function $f_k(x) = \dfrac{f(x+1/k) - f(x)}{1/k}$. The differentiability of f implies its continuity, so each f_k is continuous as well. But $\lim\limits_{k\to\infty} f_k(x) = \lim\limits_{k\to\infty} \dfrac{f(x+1/k) - f(x)}{1/k} = f'(x)$ because $1/k \to 0$ as $k \to \infty$. Hence f' is the pointwise limit of a sequence of functions from class 0 and thus belongs to class 0 (in which case it is continuous) or to class 1 (in which case it is pointwise discontinuous). Either way, derivatives must be continuous on a dense set.   Q.E.D.

We have previously seen that a differentiable function may have a discontinuous derivative, but we can now answer the big question, "How discontinuous can a derivative really be?" Thanks to Baire, the answer is, "Not very, for it must be continuous on a dense set."

Meanwhile, he continued his classification scheme:

Now suppose one has a sequence of functions belonging to classes 0 or 1 and having a limit function not belonging to either of these two classes. I will say that this limit function is of the second class, and the set of all functions that can be obtained in this manner will form class 2 [15].

To establish that there is *something* in class 2, we define a function

$$D(x) = \lim_{k\to\infty}\left[\lim_{j\to\infty}(\cos k!\,\pi x)^{2j}\right]$$

and claim that, all appearances to the contrary, this is nothing but Dirichlet's function,

$$d(x) = \begin{cases} 1 & \text{if } x \text{ is rational,} \\ 0 & \text{if } x \text{ is irrational.} \end{cases}$$

We should take a moment to verify this claim. Note first that if $x = p/q$ is a rational in lowest terms, then for any $k \geq q$, the expression $k!\,\pi x = k!\,\pi\left(\dfrac{p}{q}\right)$ is an integer multiple of π. Thus, for each k after a certain point,

$$\lim_{j\to\infty}(\cos k!\,\pi x)^{2j} = \lim_{j\to\infty}(\pm 1)^{2j} = 1, \quad \text{and so} \quad D(x) = \lim_{k\to\infty}\left[\lim_{j\to\infty}(\cos k!\,\pi x)^{2j}\right] = 1$$

as well. On the other hand, if x is irrational, then $k!\,\pi x$ cannot be an integer multiple of π, and it follows that $|\cos k!\,\pi x| < 1$. Consequently, for each k,

$$\lim_{j\to\infty}(\cos k!\,\pi x)^{2j} = 0 \quad \text{and so} \quad D(x) = \lim_{k\to\infty}\left[\lim_{j\to\infty}(\cos k!\,\pi x)^{2j}\right] = \lim_{k\to\infty} 0 = 0.$$

Because D equals 1 at each rational and 0 at each irrational, it is indeed Dirichlet's function traveling *incognito*.

What makes this intriguing is the analytic nature of D. When it was introduced early in the nineteenth century, Dirichlet's function seemed so pathological as to lie beyond the frontier of analysis. Yet here we see it as nothing worse than the double limit of some well-behaved cosines.

Moreover, for each k and j, the function $(\cos k!\,\pi x)^{2j}$ is continuous, so Dirichlet's function is seen to be the pointwise limit of the pointwise limits of continuous functions. This places it in class 0, class 1, or class 2. But we know that d is discontinuous everywhere and so does not belong to class 0 (which requires continuity) nor to class 1 (which requires continuity on a dense set). The only alternative is that Dirichlet's function resides in Baire's second class.

Baire was just getting warmed up. A function that is the pointwise limit of those from classes 0, 1, and 2 but does not belong to any of these classes is said to be in class 3. A limit of functions from classes 0, 1, 2, or 3 that escapes these will be in class 4. And on it goes. In the end we have an unimaginably vast tower of functions, beginning with continuous ones and evolving via repeated limits into ever more peculiar entities.

Needless to say, Baire's classification raised a host of questions. For instance, how can we be sure there are *any* functions in class 247? And are there functions so bizarre as to belong to no Baire class at all? It was Baire's contemporary, Henri Lebesgue, who proved that the answer to both of these questions is a resounding "yes" [16].

Although ill health brought his career to an abrupt end, René Baire carved out a share of mathematical immortality. He introduced the dichotomy between first and second category sets, proved and exploited his powerful category theorem, and provided a classification of functions that seemed to extend the boundaries of analysis to the far horizon.

As historian Thomas Hawkins observed, Baire's remarkable discoveries showed that, even at the threshold of the twentieth century, the calculus was still generating wonderful new problems [17]. In this regard, Lebesgue wrote of Baire's "rich imagination and solid critical sense" and continued,

> Baire showed us how to investigate these matters; which problems
> to pose, which notions to introduce. He taught us to consider the
> world of functions and to discern there the true analogies, the

genuine differences. In absorbing the observations that Baire made, one becomes a keen observer, learning to analyse commonplace ideas and to reduce them to notions more hidden, more subtle, but also more effective.

In the end, Lebesgue called Baire "a mathematician of the highest class," an impressive testimonial from one great analyst to another [18].

We conclude by returning to the chapter's opening passage: "Any problem relative to the theory of functions leads to certain questions relative to the theory of sets." As we have seen, Baire lived by this motto. Insofar as modern analysis has embraced his position, he deserves a large debt of gratitude.

CHAPTER 14

Lebesgue

Henri Lebesgue

As the nineteenth century became the twentieth, mathematicians had reason to congratulate themselves. The calculus had been around for over two centuries. Its foundations were no longer suspect, and many of its open questions had been resolved. Analysis had come a long way since the early days of Newton and Leibniz.

Then Henri Lebesgue (1875–1941) entered the picture. He was a brilliant doctoral student at the Sorbonne when, in 1902, he revolutionized integration theory and, by extension, real analysis itself. He did so with a dissertation that has been described as "one of the finest which any mathematician has ever written" [1].

To get a sense of his achievement, we conduct a quick review of Riemann's integral before examining Lebesgue's ingenious alternative.

RIEMANN REDUX

In previous chapters we have highlighted certain "flaws" in the Riemann integral. Some statements that mathematicians had expected to be true required additional hypotheses to render them valid. Both the fundamental theorem of calculus and the interchange of limits and integrals were false without assumptions that seemed overly restrictive.

For this latter situation, our counterexample from chapter 9 involved a sequence of functions spiking ever higher. One might argue that the limit/integral interchange failed in that situation because the functions were not uniformly bounded. But the flaw runs deeper, as is evident from the following example.

Begin with the set of the rational numbers in $[0, 1]$, which we shall denote by Q_1. Their denumerability allows us to list them as $Q_1 = \{r_1, r_2, r_3, r_4, \ldots\}$. We then define a sequence of functions

$$\phi_k(x) = \begin{cases} 1 & \text{if } x = r_1, r_2, \ldots, r_k, \\ 0 & \text{otherwise}. \end{cases}$$

Here, ϕ_k takes the value 1 at each of the first k rationals from the list and takes the value 0 elsewhere. Each such function is bounded with $|\phi_k(x)| \leq 1$, and each, equaling zero except at finitely many points, is integrable with $\int_0^1 \phi_k(x)dx = 0$.

But what about $\lim_{k \to \infty} \phi_k(x)$? Because any rational number x lies *somewhere* on the list, $\phi_k(x)$ will eventually assume, and then maintain, a value of 1 as $k \to \infty$. And, if x is irrational, $\phi_k(x) = 0$ for all k. In other words,

$$\lim_{k \to \infty} \phi_k(x) = \begin{cases} 1 & \text{if } x \text{ is rational}, \\ 0 & \text{if } x \text{ is irrational}. \end{cases} \tag{1}$$

What we have, of course, is Dirichlet's function, and so, although each ϕ_k is integrable, their pointwise limit is not. The nonintegrability of Dirichlet's function shows that, by default, $\lim_{k \to \infty} \int_0^1 \phi_k(x)dx \neq \int_0^1 \left[\lim_{k \to \infty} \phi_k(x) \right]dx$. This means that our problem with interchanging limits and integrals cannot be explained away by the unboundedness present in the example of chapter 9.

Even as these issues were being considered, there remained the question of how to characterize Riemann integrability in terms of discontinuity. In the notation of the previous chapter, mathematicians hoped to finish this sentence:

A bounded function f is Riemann integrable on $[a, b]$
if and only if D_f is _____   (2)

Everyone believed that the blank would be filled by some kind of "smallness" condition on D_f, the set of points of discontinuity. It was evident that this missing condition was not "finite" nor "denumerable" nor "first category," but its identity remained uncertain. Whoever filled in the blank by connecting continuity and Riemann integrability would make a very big splash indeed.

It was Lebesgue who settled all these scores. He did so by returning to the concepts of length and area, viewing them from a fresh perspective, and thereby providing an alternative definition of the integral. The story begins with what we now call "Lebesgue measure."

MEASURE ZERO

In a 1904 monograph, *Leçons sur l'intégration*, that grew out of his dissertation, Lebesgue described his initial goal: "I wish first of all to attach to sets numbers that will be the analogues of their lengths" [2].

He started simply enough. The length of any of the four intervals $[a, b]$, $(a, b]$, $[a, b)$, and (a, b) is $b - a$. If a set is the union of two disjoint intervals, that is, if $A = [a, b] \cup [c, d]$ where $b < c$, then we naturally let the "length" of A be $(b - a) + (d - c)$. In similar fashion, we could provide a length for any finite union of disjoint intervals.

But Lebesgue had in mind considerably more complicated sets. For instance, how should we extend the concept of length to an infinite set like $S = \left\{ 1, \frac{1}{2}, \frac{1}{3}, \frac{1}{4}, \ldots \right\}$ that we proved to be nowhere dense in chapter 13? Or how would we measure the "length" of the set of irrational numbers contained in the unit interval $[0, 1]$?

Mathematicians before Lebesgue had asked these questions. In the 1880s, Axel Harnack (1851–1888) introduced what we now call the *outer content* of a bounded set [3]. Given such a set, he began by enclosing it within a covering of finitely many intervals and using the sum of their lengths as an approximation to the set's outer content. For S above, we might consider the cover $S \subseteq \left(0, \frac{2}{7}\right) \cup \left(\frac{3}{10}, \frac{7}{10}\right) \cup \left(\frac{\pi}{4}, \frac{101}{100}\right)$, the sum of whose lengths is $\frac{2}{7} + \frac{4}{10} + \left(\frac{101}{100} - \frac{\pi}{4}\right) \approx 0.9103$.

We could refine this estimate by taking a different covering. For instance, suppose we cover S by the union of five subintervals

$$S \subseteq (0, 0.2001) \cup (0.2499, 0.2501) \cup (0.3332, 0.3334) \cup$$
$$(0.4999, 0.5001) \cup (0.9999, 1.0001).$$

Although this looks a bit strange, our strategy should be clear (see figure 14.1). The left-most interval $(0, 0.2001)$ contains all points of S except for $1/4$, $1/3$, $1/2$, and 1, and each of these has been surrounded by its own narrow interval. For this covering, the sum of the lengths is $0.2001 + 0.0002 + 0.0002 + 0.0002 + 0.0002 = 0.20009$, a much smaller number than our first value 0.9103.

At this point, Harnack advanced a bold idea: cover a bounded set E by finitely many intervals *in all possible ways*, sum the lengths of the intervals in each covering, and define the outer content $c_e(E)$ to be the limit of such sums as the length of the widest interval goes to zero.

There was much to recommend this definition. For instance, the outer content of a bounded interval turned out to be its length—exactly as one would hope. Likewise, the outer content of the single point $\{a\}$ must be zero, because for any whole number k, we can cover the set $\{a\}$ by the single interval $\left(a - \frac{1}{2k}, a + \frac{1}{2k}\right)$ of length $\frac{1}{k}$. As k grows ever larger, this length tends downward toward zero and so $c_e(\{a\}) = 0$. Again, this is as expected.

Harnack could also find the outer content of an infinite set like S. His approach is suggested by our second covering above. For any $\varepsilon > 0$, we

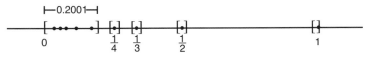

Figure 14.1

note that the interval $\left(0, \dfrac{\varepsilon}{2}\right)$ contains all but finitely many points of S,

which we denote by $\dfrac{1}{N}, \dfrac{1}{N-1}, \ldots, \dfrac{1}{2}$, and 1. We then include each of

these N points in a tiny interval of width $\dfrac{\varepsilon}{4N}$. For example, we could

place $\dfrac{1}{k}$ within $\left(\dfrac{1}{k} - \dfrac{\varepsilon}{8N}, \dfrac{1}{k} + \dfrac{\varepsilon}{8N}\right)$. Together these intervals cover S, and

the sum of their lengths is

$$\frac{\varepsilon}{2} + N\left[\frac{\varepsilon}{4N}\right] = \frac{\varepsilon}{2} + \frac{\varepsilon}{4} = \frac{3}{4}\varepsilon < \varepsilon.$$

Because, for each $\varepsilon > 0$, S lies within finitely many intervals of total length less than ε, we conclude that $c_e(S) = 0$. We have here an infinite, nowhere-dense set of zero outer content.

But Harnack confronted a different situation with the set Q_1 of rationals in $[0, 1]$: an infinite, *dense* set. He recognized that any covering of Q_1 by a finite number of intervals will of necessity cover all of $[0, 1]$. Hence $c_e(Q_1) = 1$. That is, the outer content of all rationals in the unit interval is the same as the outer content of the unit interval itself.

In some ways, this seemed to make sense, but in others it appeared problematic. For if we let I_1 be the set of *irrationals* in $[0, 1]$, identical reasoning shows that $c_e(I_1) = 1$ as well. Because the union of the disjoint sets Q_1 and I_1 is the entire interval $[0, 1]$, we see that

$$c_e(Q_1 \cup I_1) = c_e([0, 1]) = 1 \text{ yet } c_e(Q_1) + c_e(I_1) = 1 + 1 = 2.$$

Apparently, we cannot decompose a set into disjoint subsets and sum their outer contents to get the outer content of the original. Such nonadditivity was an unwelcome feature of Harnack's theory of content.

The promise of extending the concept of length to nonintervals was sufficient to lead others to modify the definition so as to eliminate the attendant problems. Many mathematicians contributed to this discussion, but history credits Lebesgue with its final resolution. He defined a set to be of *measure zero* if it "can be enclosed in a finite or a denumerable infinitude of intervals whose total length is as small as we wish" [4]. Thus a set

E is of measure zero, written $m(E) = 0$, if for any $\varepsilon > 0$, we can enclose $E \subseteq (a_1, b_1) \cup (a_2, b_2) \cup \cdots \cup (a_k, b_k) \cup \ldots$, where $\sum_{k=1}^{\infty} (b_k - a_k) < \varepsilon$. The innovation here is that Lebesgue, unlike Harnack, permitted coverings by a *denumerable infinitude* of intervals, and this made a world of difference.

It is obvious from the definitions that any subset of a set of measure zero must itself be of measure zero. It is equally clear that a set with outer content zero has measure zero as well. Thus, single points and the set S above are of measure zero. But the converse fails—and fails spectacularly—as Lebesgue showed when he proved the following.

Theorem: If a set $E = E_1 \cup E_2 \cup \cdots \cup E_k \cup \cdots$ is the denumerable union of sets of measure zero, then E is a set of measure zero also [5].

Proof: Let $\varepsilon > 0$ be given. By hypothesis, we can enclose E_1 in a denumerable collection of intervals of combined length less than $\frac{\varepsilon}{4}$ we can enclose E_2 in a denumerable collection of intervals of combined length less than $\frac{\varepsilon}{8}$, and in general we enclose E_k in a denumerable collection of intervals of length less than $\frac{\varepsilon}{2^{k+1}}$ The given set E is then a subset of the union of all these intervals which, being the denumerable union of denumerable collections, is itself a denumerable collection whose combined length is less than $\frac{\varepsilon}{4} + \frac{\varepsilon}{8} + \cdots + \frac{\varepsilon}{2^{k+1}} + \cdots = \frac{\varepsilon}{2} < \varepsilon$. Because E has been enclosed in a denumerable collection of intervals having combined length less than the arbitrarily small number ε, we see that E has measure zero.   Q.E.D.

It follows that any denumerable set is of measure zero, for such a set can be written as the (denumerable) union of its individual points. In particular, the set of rational numbers in $[0, 1]$—the dense set labeled Q_1 above—has measure zero. Because $m(Q_1) = 0$ but $c_e(Q_1) = 1$, it is evident that zero outer content and zero measure are fundamentally different.

A lesser mathematician might have retreated before the phenomenon of a dense set with measure zero. Dense sets, after all, were ubiquitous enough to be present in any interval no matter how tiny. Harnack himself had started down this path twenty years earlier but had rejected the idea as being ridiculous [6]. Such a prospect seemed sufficiently paradoxical to convince him to stick with finite coverings.

But Lebesgue was not deterred, and his approach proved its worth when he found the long-sought relationship between a function's integrability and its points of continuity. "How discontinuous can an integrable function be?" was the question. Here is the simple answer.

Theorem: For a bounded function f to be Riemann integrable on $[a, b]$, it is necessary and sufficient that the set of its points of discontinuity be of measure zero [7].

That is, he filled the critical blank in (2) with the condition $m(\mathbf{D}_f) = 0$. In many books, this is called "Lebesgue's theorem," indicating that, among the large number of results he eventually proved, this one was especially significant.

At the heart of Lebesgue's argument, not surprisingly, lay the Riemann integrability condition, which can be recast as: f is Riemann integrable if and only if, for any $\varepsilon > 0$ and any $\sigma > 0$, we can partition $[a, b]$ into finitely many subintervals in such a way that those containing points where the oscillation of the function is greater than σ (what we called the Type A subintervals) have combined length less than ε.

We observe that by the time of Lebesgue, the notion of a function's "oscillation" at a point had been made more precise than in Riemann's day. For our purposes, we shall continue to think of it informally as the maximum variability of the function in the vicinity of the point. In addition, it was known that a function is continuous at x_0 if and only if its oscillation at x_0 is zero.

Lebesgue introduced $G_1(\sigma)$ as the set of points in $[a, b]$ where the function's oscillation is greater than or equal to σ and showed that $G_1(\sigma)$ is a closed, bounded set. Because $\mathbf{C}_f = \{x \mid \text{the oscillation at } x \text{ is zero}\}$, we know that

$$\mathbf{D}_f = \{x \mid \text{the oscillation at } x \text{ is greater than zero}\}$$

$$= G_1(1) \cup G_1\left(\frac{1}{2}\right) \cup \cdots \cup G_1\left(\frac{1}{k}\right) \cup \cdots . \tag{3}$$

The validity of equation (3) should be clear. On the one hand, at any point of discontinuity, the oscillation must be positive and hence exceed $\dfrac{1}{N}$ for some whole number N. This means the discontinuity point belongs to $G_1\left(\dfrac{1}{N}\right)$ and consequently to the union on the right side of (3). Conversely, any point in this union must belong to some $G_1\left(\dfrac{1}{N}\right)$ and thus has a positive oscillation, making it a discontinuity point.

    With this background, we consider Lebesgue's argument.

Proof: First, assume the bounded function f is Riemann integrable on $[a, b]$. For any whole number k, the integrability condition guarantees that the set of points where the oscillation is greater than $\dfrac{1}{k+1}$ can be enclosed in finitely many intervals whose combined length is as small as we wish. Thus this set, as well as its subset $G_1\left(\dfrac{1}{k}\right)$, has zero content, and so $G_1\left(\dfrac{1}{k}\right)$ has measure zero. By theorem 1, the union

$$G_1(1) \cup G_1\left(\frac{1}{2}\right) \cup \cdots \cup G_1\left(\frac{1}{k}\right) \cup \cdots$$ will then be of measure zero, which implies, by (3), that D_f is of measure zero also. This completes one direction of the proof.

    For the converse, assume that $m(D_f) = 0$ and let both $\varepsilon > 0$ and $\sigma > 0$. Choose a whole number k with $\dfrac{1}{k} < \sigma$. Then the set of points where the oscillation exceeds σ is a subset of $G_1\left(\dfrac{1}{k}\right)$, which, in turn, is a subset of D_f. Hence, $G_1\left(\dfrac{1}{k}\right)$ is of measure zero and so can be enclosed in a denumerable collection of (open) intervals of total length less than ε. Because $G_1\left(\dfrac{1}{k}\right)$ is closed and bounded, Lebesgue could apply the famous Heine–Borel theorem to conclude that $G_1\left(\dfrac{1}{k}\right)$ lay within a *finite* subcollection of these open intervals [8]. This finite subcollection obviously has total length less than ε and covers not only $G_1\left(\dfrac{1}{k}\right)$ but the smaller set of points where the oscillation exceeds σ. In short, the integrability condition is satisfied and f is Riemann integrable.   Q.E.D.

Later, Lebesgue defined a property to hold *almost everywhere* if the set of points where the property fails to hold is of measure zero. With this terminology, we rephrase Lebesgue's theorem succinctly as follows: A bounded function on $[a, b]$ is Riemann integrable if and only if it is continuous almost everywhere.

We can use this characterization, for example, to give an instant proof of the integrability of the ruler function R on $[0, 1]$. As we demonstrated, R is continuous except at the set of rational points whose measure is zero. This means that the ruler function is continuous almost everywhere and so is Riemann integrable. Case closed.

Lebesgue's theorem is a classic of mathematical analysis. In light of what was to come, there is a certain irony in the fact that the person who finally understood the *Riemann* integral was the one who would soon render it obsolete: Henri Lebesgue.

THE MEASURE OF SETS

The notion of zero measure, for all of its importance, is applicable only for certain sets on the real line. As he continued his thesis, Lebesgue defined "measure" for a much larger collection of sets. The basic idea was borrowed from his countryman Emil Borel (1871–1956), but Lebesgue improved upon it (dare we say?) immeasurably.

The approach has a familiar ring. For a set $E \subseteq [a, b]$, Lebesgue wrote:

We can enclose its points within a finite or denumerably infinite number of intervals; the measure of the set of points of these intervals is . . . the sum of their lengths; this sum is an upper bound for the measure of E. The set of all such sums has a smallest limit $m_e(E)$, the *outer measure* of E [9].

Symbolically, this amounts to

$$m_e(E) = \inf\left\{\sum_{k=1}^{\infty} (b_k - a_k) \,|\, E \subseteq (a_1,b_1) \cup (a_2,b_2) \cup (a_3,b_3) \cup \cdots\right\},$$

where we have employed the *infimum*, or greatest lower bound, of the set in question. Again, the difference between outer measure and outer content is that Lebesgue allowed for denumerably infinite coverings along with the finite ones. He observed at once that $m_e(E) \le c_e(E)$, for taking more coverings can only decrease their greatest lower bound.

Next, he looked at the complement of E in [a, b] which we write as $E^c = \{x \,|\, x \in [a, b] \text{ but } x \notin E\}$. With the definition above, he found the outer measure of E^c and then defined the *inner measure* of E as $m_i(E) = (b - a) - m_e(E^c)$.

Rather than determine the inner measure of E by means of the outer measure of its complement, a modern treatment is likely to "fill" the set E from within by finite or denumerably infinite unions of intervals and then take the least upper bound, or *supremum*, of the sum of their lengths. That

is, $m_i(E) = \sup\left\{\sum_{k=1}^{\infty} (b_k - a_k) \,|\, (a_1,b_1) \cup (a_2,b_2) \cup (a_3,b_3) \cup \cdots \subseteq E\right\}$. For

bounded sets, the two approaches are equivalent, but the second one applies equally well if E is unbounded.

At this point, Lebesgue showed that "the inner measure is never greater than the outer measure," that is, $m_i(E) \le m_e(E)$, and then stated the key definition: "Sets for which the inner and outer measures are equal are called *measurable* and their measure is the common value of $m_i(E)$ and $m_e(E)$" [10].

The family of measurable sets is truly immense. It includes any interval, any open set, any closed set, and any set of measure zero, along with the set of rationals and the set of irrationals. In fact, for some time mathematicians were unable to find a set that was *not* measurable, that is, one for which $m_i(E) < m_e(E)$. These were eventually constructed by means of the axiom of choice and turned out to be extremely complicated [11].

Lebesgue explored the consequences of his definitions, three of the most basic of which were:

1. If E is measurable, then $m(E) \ge 0$.
2. The measure of an interval is its length.

3. If $E_1, E_2, \ldots, E_k, \ldots$ is a finite or denumerably infinite collection of pairwise disjoint measurable sets and if $E = E_1 \cup E_2 \cup \cdots \cup E_k \cup \cdots$ is their union, then E is measurable and $m(E) = m(E_1) + m(E_2) + \cdots + m(E_k) + \cdots$.

This third condition is the additivity property that outer content lacked. With it, we can easily find the measure of the set of irrationals in [0, 1], which we called I_1 above. We note that $[0, 1] = Q_1 \cup I_1$, where the two sets on the right are disjoint and measurable. Thus, $1 = m[0, 1] = m(Q_1 \cup I_1) = m(Q_1) + m(I_1) = 0 + m(I_1)$, and so $m(I_1) = 1$. In terms of measure, the irrationals dominate [0, 1], whereas the rationals are insignificant.

Among other things, Lebesgue measure provided a new dichotomy between "small" (measure zero) and "large" (positive measure). This took its place alongside the cardinality dichotomy (denumerable versus nondenumerable) and the topological one (first category versus second category). In all three, the rationals qualify as small for they are of measure zero, denumerable, and of the first category, whereas the irrationals are large (being of positive measure), nondenumerable, and of the second category.

To continue with this idea, we have seen that, for any of these dichotomies, subsets and denumerable unions of "small" sets are "small," and we have proved that a denumerable set is both of the first category and of measure zero. However, other "large/small" connections do not hold. It is possible to find first category sets that are nondenumerable and of positive measure and to find measure zero sets that are nondenumerable and of the second category [12]. Obviously, these concepts had carried mathematicians into some deep waters.

In his dissertation, Lebesgue was not content to consider just measurable *sets*. He defined a measurable *function* in these words: "We say that a function f, bounded or not, is measurable if, for any $\alpha < \beta$, the set $\{x \mid \alpha < f(x) < \beta\}$ is measurable" [13]. The diagram in figure 14.2 gives a geometric sense of this definition. For $\alpha < \beta$ along the y-axis, we collect all points x in the domain whose functional values fall between α and β. If this set is measurable for all choices of α and β, we say that f is a measurable function.

Using properties of measurable sets, Lebesgue showed that f is a measurable function if and only if, for any α, the set $\{x \mid \alpha < f(x)\}$ is measurable. From this result it easily follows that Dirichlet's function d is measurable, because there are only three possibilities for the set $\{x \mid \alpha < d(x)\}$: it is empty if $\alpha \geq 1$; it is the set of rationals if $0 < \alpha \leq 1$; and it is the set of all real numbers if $\alpha \leq 0$. In each case, these are measurable sets, so d is a measurable function.

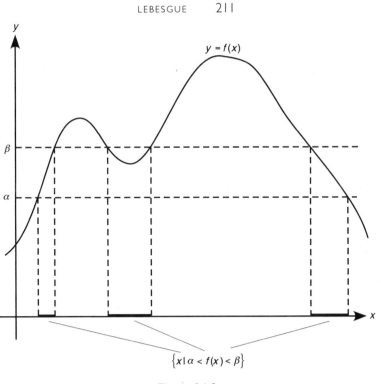

$$\{x| \alpha < f(x) < \beta\}$$

Figure 14.2

We have seen that Dirichlet's function is neither pointwise discontinuous nor Riemann integrable. With its wild behavior, it is excluded from these two families of functions. But it *is* measurable. One begins to sense that, in introducing measurable functions, Lebesgue had cast his net very widely.

He continued his line of reasoning by proving that, for a measurable function, each of the following is a measurable set:

$$\{x|f(x) = a\}, \ \{x| \alpha \le f(x) < \beta\}, \ \{x| \alpha < f(x) \le \beta\},$$
$$\text{and } \{x| \alpha \le f(x) \le \beta\}. \tag{4}$$

He also showed that sums and products of two measurable functions are measurable, implying that we cannot leave the world of measurable functions by adding or multiplying. "But," wrote Lebesgue, "there is more."

Theorem: If $\{f_k\}$ is a sequence of measurable functions and $f(x) = \lim_{k \to \infty} f_k(x)$ is their pointwise limit, then f is measurable also [14].

This is remarkable, for it says that we cannot escape the world of measurable functions even by taking pointwise limits. In (1) above we saw that this is not true of bounded, Riemann-integrable functions, and in earlier chapters we noted a similar deficiency for continuous functions or those of Baire class 1. In those situations, the family of functions was too restrictive to contain all of its pointwise limits. Measurable functions, by contrast, are strikingly inclusive.

Lebesgue was quick to observe a fascinating consequence of these theorems. We can easily see that constant functions are measurable, as is the identity $f(x) = x$. By adding and multiplying, it follows that any polynomial is measurable. The Weierstrass approximation theorem (see chapter 9) guarantees that any continuous function on $[a, b]$ is the uniform limit of a sequence of polynomials, and so any continuous function is measurable by the theorem above. For the same reason, pointwise limits of continuous functions are measurable, but these are just the functions in Baire class 1. This means that derivatives of differentiable functions are measurable. And so too are functions of Baire class 2, such as Dirichlet's function, for these are pointwise limits of functions in Baire class 1. This same reasoning reveals that any function of any Baire class is measurable.

It is fair to say that any function ever considered prior to 1900 belonged to the family of Lebesgue-measurable functions. It was a really, really big collection.

In some sense, however, all of this is prologue. Using the ideas of measure and measurable function, Lebesgue was ready to make his greatest contribution.

THE LEBESGUE INTEGRAL

Riemann's integral of a bounded function f started with a partition of the domain $[a, b]$ into tiny subintervals, built rectangles upon these subintervals whose heights were determined by the functional values, and finally let the width of the largest subinterval shrink to zero. By contrast, Lebesgue's alternative was predicated upon an idea as simple as it was bold: partition not the function's domain, but its *range*.

To illustrate, we consider the bounded, measurable function f in figure 14.3. Lebesgue let $l < L$ be the *infimum* and *supremum* of f over $[a, b]$—that is, the least upper and greatest lower bounds of the functional values—so that $[l, L]$ contained the range of the function. Then, for

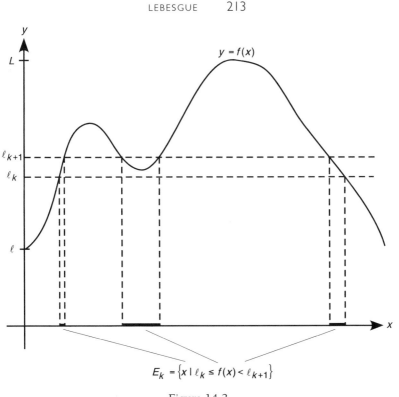

$$E_k = \left\{ x \mid \ell_k \le f(x) < \ell_{k+1} \right\}$$

Figure 14.3

any $\varepsilon > 0$, Lebesgue imagined a partition of the interval $[l, L]$ by means of the points

$$l = l_0 < l_1 < l_2 < \cdots < l_n = L,$$

where the greatest gap between adjacent partition points was less than ε.

With such a partition along the y-axis, we form the "Lebesgue sum." Like a Riemann sum, this will approximate the area under the curve with regions of known dimensions, although we can no longer be certain these regions are rectangular. Rather, we consider the subinterval $[l_k, l_{k+1})$ along the y-axis and look at the subset E_k of $[a, b]$ defined by $E_k = \{x \mid l_k \le f(x) < l_{k+1}\}$. This is the portion of the x-axis indicated in figure 14.3. Here, E_k is the union of three intervals, but its structure can be much more complicated depending on the function at hand.

At the analogous stage in Riemann's approach, we would construct a rectangle whose height was an approximation of the function's value, whose width was the length of the appropriate subinterval, and whose area was the product of these two. For Lebesgue, we use l_k to approximate the value of the function on the set E_k, but how do we determine "length" if E_k is not an interval?

The answer, which should come as no surprise, is to use the measure of the set E_k in this role. Upon multiplying height and "length," we get $l_k \cdot m(E_k)$ as the counterpart of the area of one of Riemann's thin rectangles. We sum these over all subintervals of the range to get a Lebesgue sum,

$$\sum_{k=0}^{n} l_k \cdot m(E_k),$$ where for the last term of this series we let $E_n = \{x | f(x) = l_n\}$.

Finally, Lebesgue let $\varepsilon \to 0$ so that the maximum value of $l_{k+1} - l_k$ approaches zero as well. Should this limiting process lead to a unique value, we say that f is *Lebesgue integrable* over $[a, b]$ and define

$$\int_a^b f(x)dx = \lim_{\varepsilon \to 0} \left[\sum_{k=0}^{n} l_k \cdot m(E_k) \right].$$

We must address two issues before proceeding. First, it is clear that the sets $E_0, E_1, E_2, \ldots, E_{n-1}, E_n$ partition $[a, b]$ into subsets, although not necessarily into subintervals. Second, our assumption that f is measurable implies, by (4), that each $E_k = \{x | l_k \leq f(x) < l_{k+1}\}$ along with $E_n = \{x | f(x) = l_n\}$ is a measurable set, and so we may properly talk about $m(E_k)$. Everything is falling nicely into place.

In a work written for a general audience, Lebesgue used an analogy to contrast Riemann's approach and his own [15]. He imagined a shopkeeper who, at day's end, wishes to total the receipts. One option is for the merchant to "count coins and bills at random in the order in which they came to hand." Such a merchant, whom Lebesgue called "unsystematic," would add the money in the sequence in which it was collected: a dollar, a dime, a quarter, another dollar, another dime, and so on. This is like taking functional values as they are encountered while moving from left to right across the interval $[a, b]$. With Riemann's integral, the process is "driven" by values in the domain, and values in the range fall where they may.

But, Lebesgue continued, would it not be preferable for the merchant to ignore the order in which the money arrived and instead group it by denomination? For instance, it might turn out that there were in all a dozen dimes, thirty quarters, fifty dollars, and so on. The calculation of the day's receipts would then be simple: multiply the value of the currency (which corresponds to the functional value l_k) by the number of pieces (which corresponds to the measure of E_k) and add them up. This time, as with Lebesgue's integral, the process is driven by values in the range, and the sets E_k that subdivide the domain fall where they may.

Lebesgue conceded that for the finite quantities involved in running a business, the two approaches yield the same outcome. "But for us who must add an infinite number of indivisibles," he wrote, "the difference between the two methods is of capital importance." He emphasized this difference by observing that

> our constructive definition of the integral is quite analogous to that of Riemann; but whereas Riemann divided into small subintervals the interval of variation of x, it is the interval of variation of $f(x)$ that we have subdivided [16].

To show that he was not chasing definitions pointlessly, Lebesgue proved a number of theorems about his new integral. We shall consider a few of these, albeit without proof.

Theorem 1: If f is a bounded, Riemann-integrable function on $[a, b]$, then f is Lebesgue integrable and the numerical value of $\int_a^b f(x)dx$ is the same in either case.

This is comforting, for it says that Lebesgue preserved the best of Riemann.

Theorem 2: If f is a bounded, measurable function on $[a, b]$, then its Lebesgue integral *exists*.

Here we see the power of Lebesgue's ideas, because the family of measurable functions is far more encompassing than the family of Riemann integrable ones (i.e., those continuous almost everywhere). To put it simply, Lebesgue could integrate more functions than Riemann. Theorems 1 and 2 show that Lebesgue had genuinely extended the previous theory.

For example, we have seen that Dirichlet's function is bounded and measurable on $[0, 1]$. Consequently, $\int_0^1 d(x)dx$ exists as a Lebesgue integral, in spite of the fact that it is meaningless under Riemann's theory.

Better yet, it is easy to calculate the value of this integral. We start with any partition of the range: $0 = l_0 < l_1 < l_2 < \cdots < l_n = 1$. By the nature of Dirichlet's function,

$$E_0 = \{x \mid 0 \le d(x) < l_1\} = I_1, \text{ the set of irrationals in } [0, 1],$$

$$E_k = \{x \mid l_k \le d(x) < l_{k+1}\} = \varnothing \text{ for } k = 1, 2, \ldots, n - 1,$$

$$E_n = \{x \mid d(x) = 1\} = Q_1, \text{ the set of rationals in } [0, 1].$$

For this arbitrary partition, the Lebesgue sum is

$$\sum_{k=0}^{n} l_k \cdot m(E_k) = 0 \cdot m(E_0) + l_1 \cdot m(E_1) + \cdots + l_{n-1} \cdot m(E_{n-1}) + 1 \cdot m(E_n)$$

$$= 0 \cdot m(I_1) + l_1 \cdot m(\varnothing) + \cdots + l_{n-1} \cdot m(\varnothing) + 1 \cdot m(Q_1)$$

$$= 0 \cdot 1 + l_1 \cdot 0 + \cdots + l_{n-1} \cdot 0 + 1 \cdot 0 = 0.$$

And because the Lebesgue sum is zero for *any* partition, the limit of all such is zero as well. That is, $\int_0^1 d(x)dx = 0.$

The fact that Dirichlet's function is everywhere discontinuous rendered it nonintegrable for Riemann, but such universal discontinuity was of no consequence for Lebesgue. Here was indisputable mathematical progress.

Theorem 3: If f and g are bounded, measurable functions on $[a, b]$ and $f(x) = g(x)$ almost everywhere, then $\int_a^b f(x)dx = \int_a^b g(x)dx.$

This result says that changing the values of a measurable function on a set of measure zero has no effect on the value of its Lebesgue integral. For Riemann, we can change the function's value at finitely many points without altering the integral, but once we tamper with an infinitude of points, all bets are off. By contrast, Lebesgue's integral is sufficiently tamper-proof that we can modify the function on an infinite set of zero measure yet leave the integral—and the integrability—intact.

To see this theorem in action, we revisit Dirichlet's function d and the ruler function R on $[0, 1]$ and form a trio by introducing $g(x) = 0$ for all x in $[0, 1]$. The three functions d, R, and g are certainly not identical, for they differ at rational points in the unit interval. But such differences are trivial from a measure-theoretic standpoint because $m\{x \mid d(x) \neq g(x)\} =$ $m\{x \mid R(x) \neq g(x)\} = m(Q_1) = 0$. In other words, Dirichlet's function and the ruler function equal zero almost everywhere. It follows from Theorem 3 that $\int_0^1 d(x)dx = \int_0^1 R(x)dx = \int_0^1 g(x)dx = \int_0^1 0 \cdot dx = 0$, as we have seen previously.

Yet another important result from Lebesgue's thesis is now called the bounded convergence theorem [17]. He proved that, under very mild conditions, it is permissible to interchange limits and the integral. This was a major advance over Riemann's theory.

Theorem 4 (Lebesgue's bounded convergence theorem): If $\{f_k\}$ is a sequence of measurable functions on $[a, b]$ that is uniformly bounded by the number $M > 0$ (i.e., $|f_k(x)| \leq M$ for all x in $[a, b]$ and for all $k \geq 1$) and if $f(x) = \lim_{k \to \infty} f_k(x)$ is the pointwise limit, then $\lim_{k \to \infty} \int_a^b f_k(x)dx =$ $\int_a^b f(x)dx = \int_a^b \left[\lim_{k \to \infty} f_k(x) \right] dx.$

Si les fonctions mesurables $f_n(x)$, bornées dans leur ensemble, c'est-à-dire quels que soient n et x, ont une limite $f(x)$, l'intégrale de $f_n(x)$ tend vers celle de $f(x)$.

En effet, nous savons que $f(x)$ est intégrable; évaluons

$$\int_a^b [f(x) - f_n(x)] \, dx.$$

Si l'on a toujours $|f_n(x)| < M$ et si $f - f_n$ est inférieure à ε dans E_n, $f - f_n$, étant inférieure à la fonction égale à ε dans E_n et à M dans $C(E_n)$, a une intégrale au plus égale en module à

$$\varepsilon \, m(E_n) + M \, m[C(E_n)].$$

Mais ε est quelconque, et $m[C(E_n)]$ tend vers zéro avec $\frac{1}{n}$ parce qu'il n'y a aucun point commun à tous les E_n, donc

$$\int_a^b (f - f_n) \, dx$$

tend vers zéro. La propriété est démontrée (^{1}).

Lebesgue's proof of the bounded convergence theorem (1904)

We can use this to launch our third attack upon $\int_0^1 d(x)dx$. Earlier, we introduced a sequence of functions $\{\phi_k\}$ on $[0, 1]$ for which $\lim_{k \to \infty} \phi_k(x) = d(x)$, as seen in (1). Clearly, $|\phi_k(x)| \leq 1$ for all x and all k, so this is a uniformly bounded family, and because each ϕ_k is zero except at k points, we know that each function is measurable with $\int_0^1 \phi_k(x)dx = 0$. By Lebesgue's bounded convergence theorem, we conclude yet again that

$$\int_0^1 d(x)dx = \int_0^1 \left[\lim_{k \to \infty} \phi_k(x)\right]dx$$
$$= \lim_{k \to \infty} \int_0^1 \phi_k(x)dx = \int_0^1 0 \cdot dx = 0.$$

There is time for one last flourish. We recall that Volterra had discovered a pathological function with a bounded, nonintegrable derivative. Of course, in Volterra's day, "nonintegrable" meant "non-Riemann-integrable."

By adopting Lebesgue's alternative, however, the pathology disappears. For if F is differentiable with bounded derivative F', then the Lebesgue integral $\int_a^b F'(x)dx$ must *exist* because, as we saw in chapter 13, F' belongs to Baire class 0 or Baire class 1. This is sufficient to make it Lebesgue integrable.

Better yet, the bounded convergence theorem allowed Lebesgue to prove the following [18].

Theorem 5: If F is differentiable on $[a, b]$ with bounded derivative, then

$$\int_a^b F'(x)dx = F(b) - F(a).$$

Here, back in all its original glory, is the fundamental theorem of calculus. With Lebesgue's integral, there was no longer the need to attach restrictive conditions to the derivative, for example, a requirement that it be continuous, in order for the fundamental theorem to hold. In a sense, then, Lebesgue restored this central result of calculus to a state as "natural" as it was in the era of Newton and Leibniz.

In closing, we acknowledge that many, many technicalities have been glossed over in this brief introduction to Lebesgue's work. A complete development of his ideas would require a significant investment of time and space, which makes it all the more amazing that these ideas are taken from his doctoral thesis! It is no wonder that the dissertation stands in a class by itself.

We end with a final observation from Lebesgue. In the preface of his great 1904 work, he conceded that his theorems carry us from "nice" functions into a more complicated realm, yet it is necessary to inhabit this realm in order to solve simply stated problems of historic interest. "It is for the resolution of these problems," he wrote, "and not for the love of complications, that I introduce in this book a definition of the integral more general than that of Riemann and containing it as a particular case" [19].

To resolve historic problems rather than to complicate life: a worthy principle that guided Henri Lebesgue on his mathematical journey.

Afterword

Our visit to the calculus gallery has come to an end.

Along the way, we have considered thirteen mathematicians whose careers fall into three separate periods or, at the risk of overdoing the analogy, into three separate wings.

First came the Early Wing, which featured work of the creators, Newton and Leibniz, as well as of their immediate followers: the Bernoulli Brothers and Euler. From there we visited what might be called the Classical Wing, with a large hall devoted to Cauchy and sizable rooms for Riemann, Liouville, and Weierstrass, scholars who supplied the calculus with extraordinary mathematical rigor. Finally, we entered the Modern Wing of Cantor, Volterra, Baire, and Lebesgue, who fused the precision of the classicists and the bold ideas of set theory.

Clearly, the calculus on display at tour's end was different from that with which it began. Mathematicians had gone from curves to functions, from geometry to algebra, and from intuition to cold, clear logic. The result was a subject far more sophisticated, and far more challenging, than its originators could have anticipated.

Yet central ideas at the outset remained central ideas at the end. As the book unfolded, we witnessed a continuing conversation among those mathematicians who refined the subject over two and a half centuries. In a very real sense, these creators were addressing the same issues, albeit in increasingly more complicated ways. For instance, we saw Newton expand binomials into infinite series in 1669 and Cauchy provide convergence criteria for such series in 1827. We saw Euler calculate basic differentials in 1755 and Baire identify the continuity properties of derivatives in 1899. And we saw Leibniz apply his transmutation theorem to find areas in 1691 and Lebesgue develop his beautiful theory of the integral in 1904. Mathematical echoes resounded from one era to the next, and even as things changed, the fundamental issues of calculus remained.

Our book ended with Lebesgue's thesis, but no one should conclude that research in analysis ended there as well. On the contrary, his work revitalized the subject, which has grown and developed over the past hundred years and remains a bulwark of mathematics up to the present day. That story, and the new masters who emerged in the process, must remain for another time.

We conclude as we began, with an observation from the great twentieth century mathematician John von Neumann. Because of achievements like those we have seen, von Neumann regarded calculus as the epitome of precise reasoning. His accolades, amply supported by the results of this book, will serve as the last word:

> I think it [the calculus] defines more unequivocally than anything else the inception of modern mathematics, and the system of mathematical analysis, which is its logical development, still constitutes the greatest technical advance in exact thinking. [1]

Notes

Introduction

1. John von Neumann, *Collected Works*, vol. 1, Pergamon Press, 1961, p. 3.

Chapter I
Newton

1. Richard S. Westfall, *Never at Rest*, Cambridge University Press, 1980, p. 134.

2. Ibid., p. 202.

3. Dirk Struik (ed.), *A Source Book in Mathematics, 1200–1800*, Harvard University Press, 1969, p. 286.

4. Ibid.

5. Derek Whiteside (ed.), *The Mathematical Works of Isaac Newton*, vol. 1, Johnson Reprint Corp, 1964, p. 37.

6. Ibid., p. 22.

7. Ibid., p. 20.

8. Ibid., p. 3.

9. Derek Whiteside (ed.), *Mathematical Papers of Isaac Newton*, vol. 2, Cambridge University Press, 1968, p. 206.

10. Whiteside, *Mathematical Works*, vol. 1, p. 22.

11. Ibid., p. 23.

12. Ibid.

13. Ibid., p. xiii.

14. Westfall, p. 205.

15. Whiteside, *Mathematical Works*, vol. 1, p. 4.

16. Ibid., p. 6.

17. Ibid., pp. 18–21.

18. Ibid., p. 20.

19. Whiteside (ed.), *Mathematical Papers of Isaac Newton*, vol. 2, p. 237.

20. David Bressoud, "Was Calculus Invented in India?" *The College Mathematics Journal*, vol. 33 (2002), pp. 2–13.

21. Victor Katz, *A History of Mathematics: An Introduction*, Harper-Collins, 1993, pp. 451–453.

22. C. Gerhardt (ed.), *Der Briefwechsel von Gottfried Wilhelm Leibniz mit mathematikern*, vol. 1, Mayer & Müller Berlin, 1899, p. 170.

CHAPTER 2
LEIBNIZ

1. Joseph E. Hofmann, *Leibniz in Paris: 1672–1676*, Cambridge University Press, 1974, pp. 23–25 and p. 79.

2. See, for instance, Rupert Hall, *Philosophers at War*, Cambridge University Press, 1980.

3. J. M. Child (trans.), *The Early Mathematical Manuscripts of Leibniz*, Open Court Publishing Co., 1920, p. 11.

4. Ibid., p. 12.

5. Ibid.

6. Struik, pp. 272–280, has an English translation.

7. Robert E. Moritz (ed.), *Memorabilia Mathematica*, MAA, 1914, p. 323.

8. Child, pp. 22–58.

9. Ibid., p. 150.

10. Struik, p. 276.

11. Child, p. 39.

12. Ibid., p. 42.

13. Ibid., p. 46.

14. Ibid., p. 47.

15. Ranjan Roy, "The Discovery for the Series Formula for π by Leibniz, Gregory and Nilakantha," *Mathematics Magazine*, vol. 63 (1990), no. 5, pp. 291–306.

16. Child, p. 46.

CHAPTER 3
THE BERNOULLIS

1. Howard Eves, *An Introduction to the History of Mathematics*, 5th Ed., Saunders College Publishing, 1983, p. 322.

2. Westfall, *Never at Rest*, pp. 741–743.

3. Morris Kline, *Mathematical Thought from Ancient to Modern Times*, Oxford University Press, 1972, p. 473.

4. Jakob Bernoulli, *Ars conjectandi* (Reprint), Impression anastaltique, Culture et Civilisation, Bruxelles, 1968.

5. L'Hospital, *Analyse des infiniment petits* (Reprint), ACL-Editions, Paris, 1988, pp. 145–146.

6. Struik, p. 312.

7. Dirk Struik, "The origin of l'Hospital's rule," *Mathematics Teacher*, vol. 56 (1963), p. 260.

8. Johannis Bernoulli, *Opera omnia*, vol. 3, Georg Olms, Hildesheim, 1968, pp. 385–563.

9. The *Tractatus* is appended to Jakob Bernoulli's *Ars conjectandi* (cited above), pp. 241–306.

10. See William Dunham, *Journey through Genius*, Wiley, 1990, pp. 202–205.

11. Hofmann, p. 33.

12. Jakob Bernoulli, *Ars conjectandi*, p. 250.

13. Ibid., p. 251.

14. Ibid., p. 252.

15. Ibid., pp. 246–249.

16. Ibid., p. 254.

17. Johannis Bernoulli, *Opera omnia*, vol 1, Georg Olms, Hildesheim, 1968, p. 183.

18. Johannis Bernoulli, *Opera omnia*, vol. 3, p. 388.

19. Ibid., p. 376.

20. Johannis Bernoulli, *Opera omnia*, vol. 1, pp. 184–185.

21. Johannis Bernoulli, *Opera omnia*, vol. 3, pp. 376–381.

22. Ibid., p. 381.

23. Ibid., p. 377.

CHAPTER 4
EULER

1. Eric Temple Bell, *Men of Mathematics*, Simon & Schuster, 1937, p. 139.

2. Leonhard Euler, *Foundations of Differential Calculus*, trans. John Blanton, Springer-Verlag, 2000.

3. Ibid., p. 51.

4. Ibid., p. 52.

5. Ibid.

6. Ibid., p. 116.

7. These integrals appear, respectively, in Leonhard Euler's, *Opera omnia*, ser. 1, vol. 17, p. 407, *Opera omnia*, ser. 1, vol. 19, p. 227, and *Opera omnia*, ser. 1, vol. 18, p. 8.

8. Euler, *Opera omnia*, ser. 1, vol. 18, p. 4.

9. T. L. Heath, (ed.), *The Works of Archimedes*, Dover, 1953, p. 93.

10. Howard Eves, *An Introduction to the History of Mathematics*, 5th Ed., Saunders, 1983, p. 86.

11. Euler, *Opera omnia*, ser. 1, vol. 16B, p. 3.

12. Ibid., pp. 14–16.

13. Ibid., p. 277.

14. Leonhard Euler, *Introduction to Analysis of the Infinite*, Book 1, trans. John Blanton, Springer-Verlag, 1988, p. 137.

15. Euler, *Opera omnia*, ser. 1, vol. 6, pp. 23–25.

16. Euler, *Introduction to Analysis of the Infinite*, Book I, pp. 142–146.

17. Ivor Grattan-Guinness, *The Development of the Foundations of Mathematical Analysis from Euler to Riemann*, MIT Press, 1970, p. 70.

18. Euler, *Opera omnia*, ser. 1, vol. 10, p. 616.

19. Euler, *Opera omnia*, ser. 1, vol. 4, p. 145.

20. Philip Davis, "Leonhard Euler's Integral," *American Mathematical Monthly*, vol. 66 (1959), p. 851.

21. P. H. Fuss (ed.), *Correspondance mathématique et physique*, The Sources of Science, No. 35, Johnson Reprint Corp., 1968, p. 3.

22. Euler, *Opera omnia*, ser. 1, vol. 14, p. 3.

23. Ibid., p. 13.

24. Euler, *Opera omnia*, ser. 1, vol. 16A, p. 154.

25. Ibid., p. 155.

26. Euler, *Opera omnia*, ser. 1, vol. 18, p. 217.

27. John von Neumann, *Collected Works*, vol. 1, Pergamon Press, 1961, p. 5.

CHAPTER 5
FIRST INTERLUDE

1. Struik, p. 300.

2. George Berkeley, *The Works of George Berkeley*, vol. 4, Nelson & Sons, London, 1951, p. 53.

3. Ibid., p. 67.

4. Ibid., p. 89.

5. Ibid., p. 68.

6. Ibid., p. 77.

7. Ibid., p. 72.

8. Ibid., p. 73.

9. Ibid., p. 74.

10. Carl Boyer, *The Concepts of the Calculus*, Hafner, 1949, p. 248.

11. Struik, p. 344.

12. Joseph-Louis Lagrange, *Oeuvres*, vol. 9, Paris, 1813, p. 11 (title page).

13. Ibid., pp. 21–22.

14. Augustin-Louis Cauchy, *Oeuvres*, ser. 2, vol. 2, Paris, pp. 276–278.

15. Judith Grabiner, *The Origins of Cauchy's Rigorous Calculus*, MIT Press, 1981, p. 39.

16. Berkeley, p. 76.

CHAPTER 6
CAUCHY

1. Bell, p. 292.

2. The *Cours d'analyse* is available in Cauchy's *Oeuvres*, ser. 2, vol. 3, and the *Calcul infinitésimal* appears in the *Oeuvres*, ser. 2, vol. 4.

3. Cauchy, *Oeuvres*, ser. 2, vol. 4, in the *Advertisement* of the *Calcul infinitésimal*.

4. Ibid., p. 13.

5. Ibid., p. 16.

6. Ibid., p. 20.

7. Ibid., p. 19.

8. Ibid., p. 23.

9. Kline, p. 947.

10. Cauchy, *Oeuvres*, ser. 2, vol. 3, pp. 378–380. Or see Judith Grabiner, *The Origins of Cauchy's Rigorous Calculus*, MIT Press, 1981, pp. 167–168 for an English translation.

11. Grabiner, p. 69.

12. Cauchy, *Oeuvres*, ser. 2, vol. 4, pp. 44–46.

13. Euler, *Opera omnia*, ser. 1, vol. 11, p. 5.

14. Cauchy's theory of the integral is taken from his *Oeuvres*, ser. 2, vol. 4, pp. 122–127.

15. Ibid., pp. 151–155.

16. Ibid., p. 220.

17. Ibid., pp. 226–227.

18. Cauchy, *Oeuvres*, ser. 2, vol. 3, p. 123.

19. Ibid., p. 137–138.

20. Boyer, p. 271.

CHAPTER 7
RIEMANN

1. Euler, *Introduction to Analysis of the Infinite*, Book 1, pp. 2–3.

2. Euler, *Foundations of Differential Calculus*, p. vi.

3. Israel Kleiner, "Evolution of the Function Concept: A Brief Survey," *College Mathematics Journal*, vol. 20 (1989), pp. 282–300.

4. Thomas Hawkins, *Lebesgue's Theory of Integration*, Chelsea, 1975, pp. 3–8.

5. Ibid., pp. 5–6.

6. G. Lejeune Dirichlet, *Werke*, vol. 1, Georg Riemer Verlag, 1889, p. 120.

7. Ibid., pp. 131–132.

8. Hawkins, p. 16.

9. Bernhard Riemann, *Gesammelte Mathematische Werke*, Springer-Verlag, 1990, p. 271.

10. Ibid.

11. Ibid., p. 274.

12. Ibid.

13. Ibid., p. 270.

14. Dirichlet, *Werke*, vol. 1, p. 318.

15. Riemann, p. 267.

CHAPTER 8
LIOUVILLE

1. Struik, p. 276.

2. Euler, *Introduction to Analysis of the Infinite*, Book I, p. 4.

3. Ibid., p. 80.

4. Kline, pp. 459–460.

5. Ibid., p. 593.

6. These and other aspects of Liouville's career are treated in Jesper Lützen's scientific biography, *Joseph Liouville 1809–1882: Master of Pure and Applied Mathematics*, Springer-Verlag, 1990.

7. E. Hairer and G. Wanner, *Analysis by Its History*, Springer-Verlag, 1996, p. 125.

8. J. Liouville, "Sur des classes très-étendues de quantités dont la valeur n'est ni algébrique, ni même réductible à des irrationnelles algébriques," *Journal de mathématiques pures et appliquées*, vol. 16 (1851), pp. 133–142.

9. This is adapted from George Simmons, *Calculus Gems*, McGraw-Hill, 1992, pp. 288–289.

10. Liouville, Ibid.

11. Ibid., p. 140.

12. Lützen, pp. 79–81.

13. Bell, p. 463.

14. See, for instance, the discussion in Dunham, pp. 24–26.

15. Andrei Shidlovskii, *Transcendental Numbers*, de Gruyter, 1989, p. 442.

CHAPTER 9
WEIERSTRASS

1. This biographical sketch is drawn from the Weierstrass entry in the *Dictionary of Scientific Biography*, vol. XIV, C.C. Gillispie, editor-in-chief, Scribner, 1976, pp. 219–224.

2. Bell, p. 406.

3. Hairer and Wanner, p. 215.

4. Cauchy's *Oeuvres*, ser. 2, vol. 3, p. 120.

5. See Hawkins, p. 22.

6. Victor Katz, *A History of Mathematics: An Introduction*, Harper-Collins, 1993, p. 657.

7. Hawkins, pp. 43–44.

8. Karl Weierstrass, *Mathematische Werke*, vol. 2, Berlin, 1895, pp. 71–74.

9. Quoted in Kline, p. 973.

10. Hairer and Wanner, p. 261.

11. Kline, p. 1040.

CHAPTER 10
SECOND INTERLUDE

1. Johannes Karl Thomae, *Einleitung in die Theorie der bestimmten Integrale*, Halle, 1875, p. 14.

2. Hairer and Wanner, p. 219.

3. Hawkins, p. 34.

CHAPTER 11
CANTOR

1. Georg Cantor, *Gesammelte Abhandlungen*, Georg Olms Hildesheim, 1962, p. 182.

2. Joseph Dauben, *Georg Cantor: His Mathematics and Philosophy of the Infinite*, Princeton University Press, 1979, p. 1.

3. Ibid., p. 136.

4. Cantor, pp. 115–118.

5. Dauben, p. 45.

6. Ibid., p. 49.

7. Cantor, p. 278.

8. Ibid., p. 116.

9. Bertrand Russell, *The Autobiography of Bertrand Russell*, vol. 1, Allen and Unwin, 1967, p. 127.

10. Bell, p. 569.

11. Russell, p. 217.

CHAPTER 12
VOLTERRA

1. This biographical sketch is based on the Volterra entry in the *Dictionary of Scientific Biography*, vol. XIV, pp. 85–87.

2. Vito Volterra, *Opere matematiche*, vol. 1, Accademia Nazionale dei Lincei, 1954, pp. 16–48.

3. Hawkins, pp. 56–57.

4. Ibid., p. 30.

5. H. J. S. Smith, "On the Integration of Discontinuous Functions," *Proceedings of the London Mathematical Society*, vol. 6 (1875), p. 149.

6. Hawkins, pp. 37–40.

7. Volterra, pp. 7–8.

8. Ibid., p. 8.

9. Ibid., p. 9.

10. Kline, p. 1023.

CHAPTER 13
BAIRE

1. René Baire, *Sur les fonctions des variables réelles*, Imprimerie Bernardoni de C. Rebeschini & Co., 1899, p. 121.

2. This information is taken from the *Dictionary of Scientific Biography*, vol. I, Scribner, 1970, pp. 406–408.

3. Adolphe Buhl, "René Baire," *L'enseignment mathématique*, vol. 31 (1932), p. 5.

4. See Hawkins, p. 30.

5. Baire, p. 65.

6. Ibid.

7. Ibid.

8. Ibid., p. 66.

9. Ibid., pp. 64–65.

10. Ibid., p. 66.

11. Ibid.

12. Ibid., pp. 66–67.

13. Ibid., p. 68.

14. Baire, pp. 63–64 or, for a modern treatment, see Russell Gordon, *Real Analysis: A First Course*, Addison-Wesley, 1997, pp. 254–256.

15. Ibid., p. 68.

16. Henri Lebesgue, "Sur les fonctions représentables analytiquement," *Journal de mathématiques*, (6), vol. 1 (1905), pp. 139–216.

17. Hawkins, p. 118.

18. Lebesgue's quotations appear in "Notice sur René-Louis Baire" from the *Comptes rendus des séances de l'Académie des Sciences*, vol. CXCV (1932), pp. 86–88.

CHAPTER 14
LEBESGUE

1. Quoted in G.T.Q. Hoare and N. J. Lord, "'Intégrale, longueur, aire'—the centenary of the Lebesgue integral," *The Mathematical Gazette*, vol. 86 (2002), p. 3.

2. Henri Lebesgue, *Leçons sur l'intégration et la recherché des fonctions primitives*, Gauthier-Villars, 1904, p. 36.

3. Hawkins, p. 63.

4. Lebesgue, p. 28

5. Ibid.

6. Hawkins, p. 64.

7. Lebesgue, pp. 28–29.

8. The Heine–Borel theorem for closed, bounded sets of real numbers is a staple of any analysis text; see, for instance, Frank Burk's *Lebesgue Measure and Integration*, Wiley, 1998, p. 65. Its history is intricate, but we note that Lebesgue's thesis contains a beautiful proof on pp. 104–105, yet another highlight of his remarkable dissertation. For more information, see Pierre Dugac, "Sur la correspondance de Borel et le théoreme de Dirichlet-Heine-Weierstrass-Borel-Schoenflies-Lebesgue," *Archives internationales d'histoire des sciences*, 39 (122) (1989), pp. 69–110.

9. Lebesgue, p. 104.

10. Ibid., p. 106.

11. Burk, pp. 266–272.

12. See Bernard Gelbaum and John Olmsted, *Counterexamples in Analysis*, Holden-Day, 1964, p. 99.

13. Lebesgue, p. 111.

14. Ibid.

15. Henri Lebesgue, *Measure and the Integral*, Holden-Day, 1966, pp. 181–182.

16. Henri Lebesgue, *Leçons sur l'intégration et la recherché des fonctions primitives*, AMS Chelsea Publishing, 2000, p. 136. (This is a reprint of the second edition, originally published in 1928, of Lebesgue's 1904 work that we have been citing above.)

17. Lebesgue, 1904, p. 114.

18. Ibid., p. 120.

19. Ibid., pp. v–vi.

AFTERWORD

1. John von Neumann, *Collected Works*, vol. 1, Pergamon Press, 1961, p. 3.

Index